Aerobic Granule Reactor Technology

STOWA, the Dutch Foundation for Applied Water Research, is an organisation for the initiation, co-ordination and application of applied research for the benefit of all water authorities responsible for water management in The Netherlands. STOWA produces scientific publications in the applied fields of wastewater treatment, urban water management, sludge disposal, surface and groundwater quantity and quality, aquatic ecology and flood protection. The water authorities in the foundation provide the finances needed to carry out the research programme and to run the STOWA secretariat.

STOWA REPORT

Aerobic Granule Reactor Technology

M.K. de Kreuk

L.M.M. de Bruin

IWA Publishing

stowa
Foundation for Applied Water Research

Published by IWA Publishing, Alliance House, 12 Caxton Street, London SW1H 0QS, UK
Telephone: +44 (0) 20 7654 5500; Fax: +44 (0) 20 7654 5555; Email: publications@iwap.co.uk
Web: www.iwapublishing.com

First published 2004

Originally published as *Development of Aerobic Granule Reactor Technology*. STOWA report 2003-07, STOWA, Utrecht, the Netherlands, ISBN 90 5773 2106.

British Library Cataloging in Publication Data
A CIP catalogue record for this book is available from the British Library

Library of Congress Cataloging- in-Publication Data
A catalog record for this book is available from the Library of Congress

ISBN: 1 84339 067 1

STOWA Project Implementation

Merle de Kreuk (TUD), Mark van Loosdrecht (TUD), Bart de Bruin (DHV), Helle van der Roest (DHV), Joost van der Pluijm (van der Pluijm Water- en Milieumanagement)

STOWA Advisory Committee

Tom Wouda (Waterschap De Aa; vz), Jan Willem Mulder (Zuiveringsschap Hollandse Eilanden en Waarden), Frank Brandse (Waterschap Reest & Wieden), Hardy Temmink (Wageningen Universiteit), Wim Wiegant (Royal Haskoning), Paul Roeleveld, later opgevolgd door Cora Uijterlinde (STOWA).

TABLE OF CONTENTS

PREFACE

A new development for the treatment of domestic wastewater is a technology based on aerobic granular sludge. Granular sludge can be developed under specific process conditions. Because of the unique properties of this granular sludge, high volumetric loading rates of aeration tanks can be applied. Thanks to excellent settling properties, the separation of treated wastewater and granular sludge can take place at high hydraulic loading rates.

Depending on the chosen process configuration good effluent quality can be obtained, complying with the stringent effluent requirements regarding nitrogen and phosphorus, which can be expected in the future. In this way aerobic granular sludge has the potential to contribute significantly to wastewater treatment management.

During this research, the possibility of achieving high efficiency of biological removal of nitrogen and phosphorus was tested. The technical and economical feasibility of the technology was also assessed. It was possible to sustain stable granules for a longer period and regarding COD, total nitrogen and phosphate high removal efficiencies were obtained. Further, in comparison with conventional treatment techniques, WWTP based on granular sludge is cheaper and needs significantly less volume.

Taking into account these results as well as the implementation in practice, STOWA decided to continue the development of this technology in the form of pilot-research.
This report describes the results of the research. The research was performed in 3 litre reactors in the laboratory of the Technical University in Delft.

Utrecht, June 2003

De directeur van de STOWA
ir. J.M.J. Leenen

SUMMARY

Conventional sewage treatment plants (STPs) based on activated sludge technology require a large footprint. This is caused by the relatively poor settling characteristics of activated sludge, resulting in low permissible dry solids (DS) concentrations in aeration tanks and in a low maximum hydraulic loads of secondary sedimentation tanks. In the nineties of last century, compact attached growth technologies in several configurations were developed (immobilised bed, fluidised bed and airlift reactors). Main feature of these continuously operated reactors is a high volumetric load, occasionally without a separate sludge/water separation step resulting in a small footprint. Recent laboratory research showed that in a discontinuously fed system, dense, more stable and smooth granular sludge can be grown without a carrier. The discontinuous feeding pattern also allows an efficient nutrient removal.

During recent laboratory research stable granulation was obtained with a three-minute pulse feeding period (total cycle time 3 hours), a synthetic influent (acetate ammonium mixture) and a high dissolved oxygen (DO) concentration. At a low DO (40% of saturation concentration) and a longer feeding period the granules became unstable and fell apart. It also was not possible to maintain granules under these conditions in a bubble column reactor. However, for a full-scale application of the aerobic granular sludge technology it is essential to allow longer feeding periods and operation at low DO. For economic reasons a bubble column is preferred to an airlift reactor. Therefore, this project aimed at elimination of these bottlenecks for scale-up and at evaluation of the economic and technical feasibility of the granular sludge sequencing batch reactor (GSBR). Laboratory research focussed mainly on the possibilities for simultaneous nutrient removal. Therefore, the period and way of feeding was varied and granulation at low oxygen concentrations was studied. The influence of the use of a bubble column on granulation was experimentally determined as well. Finally, formation of aerobic granular sludge with pre-treated sewage was investigated.

In order to provoke a "pulse feeding" it was proposed to feed during 33 % of the cycle time under anaerobic conditions. If substrate is not converted during this period, substrate concentrations would be similar to a pulse feeding operation when aeration starts. Moreover, feeding influent in a upflow mode through the granular sludge blanket will allow biological phosphate removal bacteria to proliferate. In this case, substrate is taken-up in the cells during the anaerobic feeding period and is cell-internally stored as a polymer. It was shown that a 60 minutes feeding period of synthetic influent (total cycle time 3 hours) and a DO of 20%, caused a simultaneous COD (Acetate), phosphate and total nitrogen removal, (100% COD (acetate) and P removal, 100% ammonia removal by nitrification and 90% total N-removal). Biomass concentrations that can be maintained in this type of reactors was around 5 times higher than in an activated sludge system. This shows the potential of the technology.

In a separate experiment granulation on pre-treated (pre-settled) sewage as carbon source was tested. This experiment was successful in showing granulation. All biodegradable dissolved COD was removed from the influent. The possibilities of extensive nutrient removal from municipal wastewater will have to be investigated in follow-up research.

The feasibility study showed that the aerobic granular sludge technology seems very promising. Based on total annual costs both GSBR variants (GSBR with pre-treatment and GSBR with post-treatment) prove to be more attractive than the reference activated sludge alternatives (6-16%). From a sensitivity analysis it appears that the GSBR technology is less sensitive to the land price and more sensitive to a higher RWF. This means that the GSBR technology becomes more attractive at lower RWF/DWF ratios and higher land prices. Because of the high permissi-

ble volumetric load the footprint of the GSBR variants is only 25% compared to the references. However, the GSBR with only primary treatment cannot meet the present effluent standards for municipal wastewater in the Netherlands, mainly because of an expected too high suspended solids concentration in the effluent.
A growing number of sewage treatment plants in the Netherlands is going to be faced with more stringent effluent standards. In general, activated sludge plants will have to be extended with a post treatment step (e.g. sand filtration) or be transformed into a Membrane Bio Reactor. In this case a GSBR variant with primary treatment as well as post treatment can be an attractive alternative.

Based on both the good research results and the positive technical and financial evaluation for full-scale application for sewage treatment, it is recommended to develop the aerobic sludge technology further by means of pilot research at a sewage treatment plant.

1 INTRODUCTION

Most wastewater treatment systems have the disadvantage of large area requirements and high excess biomass production. To build compact reactors, sludge can be grown in biofilms on carrier particles with high settling rates. The process conditions of aerated fluidised bed reactors are simple and the area requirement is small, because of the large specific biofilm area. Most of these compact biofilm reactors are continuously operated systems. An integrated settler on top of the reactor keeps the biofilm-covered particles in the reactor. Recent research showed the advantages of a discontinuously fed system and the possibility to grow stable granular sludge under aerobic circumstances (Beun, 1999). Full biological N- and P-removal can be easily integrated in a discontinuous system. Because of the high settling capacity of granules, the use of a traditional settler is not necessary and therefore, the installation can be built very compact.

History of the project

In previous research, a sequenced batch airlift reactor was used in order to obtain aerobic granular sludge (Beun, 2001). The used cycle configuration consisted of a pulse feed of the nutrient mixture, a long aeration period, a short settling period and an effluent withdrawal period. The total cycle time was three hours.

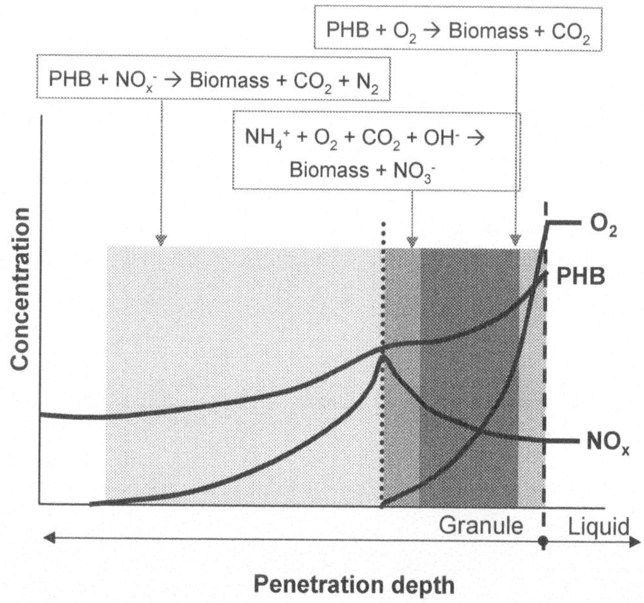

Figure 1 Schematic concentration profiles of the different compounds in the granule and their conversion processes (PHB: Poly-β-hydroxybutyrate).

The first advantage of a discontinuous fed reactor is the possibility to select for particles with high settling velocities. This selection can be controlled through the duration of the settling phase. The particles which settle fast enough will be kept in the reactor, the slow settling floc structures will be washed out with the effluent. The second advantage is the high pulse COD concentration in the beginning of the cycle. Part of the COD will be used for aerobic growth, and part will diffuse into the granule and will be converted into poly-β-hydroxybutyrate (PHB) (feast period; period in which cell external substrate is available). When the COD in the solution is consumed, the organisms will use the slowly degradable storage product PHB as a carbon source for growth (famine period; starvation period in which cell external substrate is not available). The slowly degradable PHB leads to a low growth rate and therefore to a dense and smooth biofilm (Picioreanu, 1999). In the aerobic outer layer of a granule, PHB is converted with oxygen and nitrification occurs. Because of these processes, oxygen does not reach the inner part of the granule, in which nitrate and PHB will be present, leading to denitrification (figure 1). Preliminary modelling of this system showed that a dissolved oxygen concentration of 40% of air saturation would lead to a maximum nitrogen removal. Higher oxygen concentrations will lead to nitrate in the effluent, lower concentrations to ammonia (Beun et al., 2001).

Objectives

The main objective of this project was the development of the granule reactor technology, referring to optimal process conditions as well as the final technical design. The scale-up from laboratory to full-scale conditions played a central role during this research. The formation and the stability of granules under different circumstances were investigated. The circumstances were changed according to the scale-up criteria that followed from technical and economical design. Furthermore, the study of conversion processes in granular sludge was continued.
The research focused on the following subjects:

- Repetition of the granulation experiments and increasing the understanding of granulation;
- Effect of different feeding patterns and cycle configuration;
- Effect of DO concentration on granulation and conversion processes;
- Effect of the shear stress in the reactor (airlift or bubble column);
- Design and cost-calculations of a full-scale wastewater treatment plant based on aerobic granules.

In chapter 2 of this report, the used methods are described. Chapter 3 describes the different feeding patterns with a synthetic influent and their influence on the stability of the granules and the conversion processes in the granules. In chapter 4 the effect of varying oxygen concentrations will be presented with two different feeding patterns. In chapter 5 granulation in airlift reactors and bubble columns are compared applying a long feeding period with synthetic influent. In chapter 6 granulation on municipal sewage is described and in chapter 7 the economic and design part of the study is specified. Chapter 8 contains overall conclusions and recommendations.

2 METHODS

Reactor set-up

The experimental set-up is given in Figure 2. The reactors have a 3-litres working volume (diameter 6.25 cm; filled reactor height ±102 cm, total reactor height 130 cm), are double walled, contain an inner tube (in case of the sequencing batch airlift reactors (SBAR)) and an aerator to blow in the air at 4 l/s (= 87.5 m/hr). This airflow causes enough pressure difference between outer and inner tube to move the liquid and the granules. In case of the sequencing batch bubble column (SBBC) a larger sparger is used to distribute the gas bubbles more evenly and to obtain suspension of the particles. Effluent is extracted at a height of 50 cm, which means that after every cycle 1.4 litres remains in the reactor (an exchange ratio ($V_{effluent}/V_{filled\ reactor}$) of approximately 0.53).

Operational procedures

At start-up, the reactor was inoculated with activated sludge from sewage treatment plant (STP) Gouda. When a quick development of Bio-P activity was desired, sludge from STP Hardenberg was added as well, because of its high activity of phosphate accumulating organisms (PAO). During the start-up with pre-treated sewage (excessive pre-settling in the lab), the reactor was also inoculated with a small amount (<25 mg VSS) of crushed granules from the reactors operated with synthetic medium, in order to get a desired biomass composition, but to avoid the influence of granules on the overall granulation process.

The arrangement of the cycle is given in figure 3. The total cycle time was 3 hours and 2 hours in the experiment with pre-treated wastewater. A cycle consisted of a fill, aeration, settling and effluent withdrawal phase. The idle phase is not applied in the experiments. The reactors had the possibility to apply different filling regimes (figure 3). The influent was filled from the top (set-up 1) or from the bottom through the settled particle bed (set-up 2 and 3). During the aeration periods with controlled oxygen concentrations, air could be reused via an off-gas circulation and a membrane pump (4 l/s). The system could also be operated without off-gas circulation and by applying pressurised air for high uncontrolled oxygen levels. The settling time (3 minutes) was chosen such that only particles with a settling velocity larger than 10 m/h were effectively retained in the reactor. In all systems, feeding was anaerobic and influent was either quickly fed from the top (first feeding pattern) or during a longer period from the bottom of the reactor and flowed in a plug-flow pattern through the bed (second feeding pattern).

The composition of the synthetic influent is given in table 1. The composition and selection of the pre-treated sewage water, which is used during the experiments, is given in chapter 6.

Table 1 Composition of the dosed media

Medium	component	final concentration in influent	remarks
Medium A:	NaAc	371.5 mg COD/l	COD-load = 1.6 kg/m^3/day
	$MgSO_4 \cdot 7H_2O$	82.6 mg/l	
	KCl	32.5 mg/l	
Medium B:	NH_4Cl	46.9 mg NH_4-N/l	COD/N ratio = 7.9
	K_2HPO_4	12.5 mg PO_4-P/l	P-concentration = 20 mg P/l
	KH_2PO_4	6.3 mg PO_4-P/l	
	Trace element solution	0.93 ml/l	
Water:	Tap-water		total influent 1600 ml per cycle

An overview of the experiments that are carried out is given in table 2. The operation time of the reactor under different circumstances was enough to reach a steady state if possible and operate the reactor for at least 3-4 sludge ages under steady state conditions.

Table 2 Overview of the laboratory experiments

	Experiment	Reactor	Operation time	Remarks
1a	Pulse feed, 100% DO (Stage I)	SBAR 1	143 days	Same experimental set-up as Beun
1b	Pulse feed, 40% DO (Stage II)	SBAR 1	33 days	Continued from 1a
1c	Pulse feed, 20% DO (Stage III)	SBAR 1	30 days	New start-up with low DO
2a	Long feed, 100% DO	SBAR 2	233 days	First 103 days with nitrification inhibition by dosing ATU
2b	Long feed, 40% DO	SBAR 2	184 days	Continued from 2a
2c	Long feed, 20% DO	SBAR 2	250 days	Continued from 2b
3	Long feed, 40% DO	SBBC 1	385 days	
4	Pretreated sewage	SBAR 3	105ays	Details in table 6, chapter 6

Control
The pH and oxygen concentration were measured continuously. The pH was controlled between pH 6.8 and 7.2 by the pH electrode and dosage of 1M NaOH or 1M HCl. The temperature was kept at 20 °C.
The DO concentration in the liquid was measured and controlled by dosing extra air or nitrogen gas in the off gas recycle flow via mass flow controllers (mfc, figure 2). This led to a stable DO in the reactor.

Analytical procedures
The CO_2 concentration in the off-gas was measured online with an infrared CO_2 analyser. Dry weight (TSS) in the reactor was determined by filtering and drying a sample for at least 24 hours at 105°C. Ash content was determined according to the Dutch standard method NEN6621 (NNI, 1982). Effluent biomass concentration was determined by the measured VSS (dry weight - ash content).
Solid retention time (SRT) is determined from the effluent biomass, which consists of granules and eroded material from the granule surface. Applying this method can lead to an underestimation of the sludge age in the inner part of the granule.
Bed volume of the settled bed was determined by reading the height of the biomass bed directly from the scale on the reactor at the end of a settling and effluent withdrawal period (in total 8 minutes settling time). Sludge volume index (SVI) was determined with the volume of the settled bed and the TSS concentration in the reactor. This unconventional method of SVI measurement was chosen to minimise the disturbance of the granules by leaving them in the reactor contrary to a conventional SVI measurement. Because of the compact structure of the granules, the SVI after 8 minutes of settling is equal to the SVI after an hour. Therefore, the SVI values obtained in this research can easily be compared with the ones obtained in activated sludge systems.

Concentrations of PO_4^{3-}, NO_3^-, NO_2^-, NH_4^+ and Acetate were measured regularly in filtered effluent samples and in filtered contents of the reactor after a filling period and occasionally

every 10 to 20 minutes during one cycle (cycle measurement). The PHB fraction of the biomass was measured during the cycle measurements too. The influent concentration was calculated from the known medium concentration and additions. PO_4^{3-} and NH_4^+ concentrations were measured with Dr. Lange test kits. NO_3^- and NO_2^- concentrations were regularly checked with Merck indicator papers and measured spectrophotometrically. Acetate concentration of the filtered samples and PHB of freeze-dried granules were measured by gas chromatography.
The elemental composition of the granules (C, H, N, S) was measured by flash combustion in a partial oxygen atmosphere at 1,020°C using a Carlo Erba EA1108 elemental analyser.

Changes in morphology of the granules were followed by image analyses (IA), with a Lexmark Optra image analysis system. The average equivalent diameter (d) was measured, as well as the capriciousness of the particle surface (shape factor, 0=line, 1=circle) and the roundness of the particle (aspect ratio, 0=line, 1=circle). Also pictures were taken with this system to follow the development of the granules.

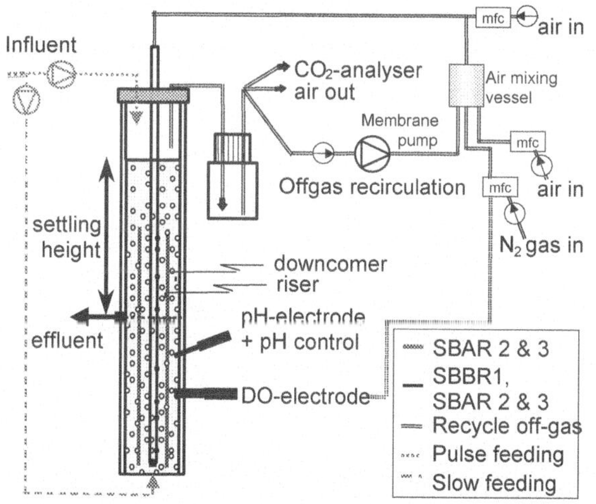

Figure 2 Set up of the laboratory scale reactors

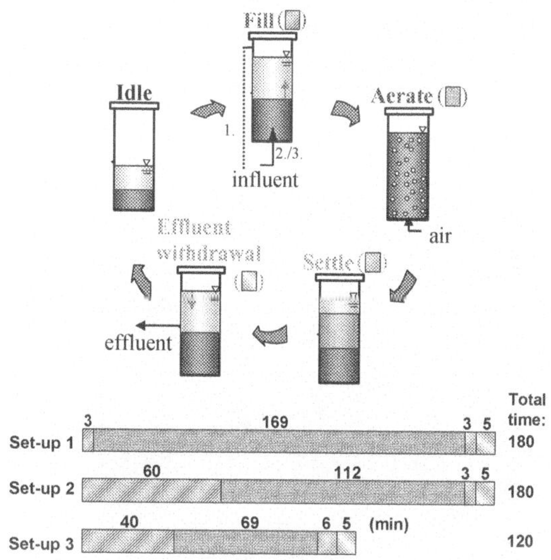

Figure 3 The cycle configuration and different cycle times for the various experiments

3 INFLUENCE OF FEEDING TIME

3.1 Introduction

In previous research (Beun, 2001, Etterer and Wilderer, 2001), aerobic granules were successfully grown in an airlift reactor with an anaerobic filling time of 3 minutes. This 3 minutes pulse feed was supposed to be a necessity for the development of smooth and dense particles. When aeration starts after the pulse feed and at the beginning of the aerobic period, the acetate concentration in the liquid is high enough to penetrate throughout the whole granule and to be partially converted in the slowly biodegradable storage polymer polyhydroxy-β-butyrate, or PHB (feast period). When all acetate is consumed, partly directly for growth (10-40%), partly for PHB production (60-90%), the organisms will use the stored PHB for growth and maintenance (famine period) (Beun et al., 2000). The conversion rate of PHB by heterotrophic bacteria is much lower than the conversion rate of easy biodegradable acetate. From earlier biofilm research it is known that a low conversion rate and thus low actual growth rates lead to denser and smoother biofilms than at a high conversion rate (Picioreanu, 1999, Van Loosdrecht, 1995). Formation of slowly degradable PHB out of rapidly degradable acetate is one of the most essential mechanisms for the production of dense, smooth and stable granules.

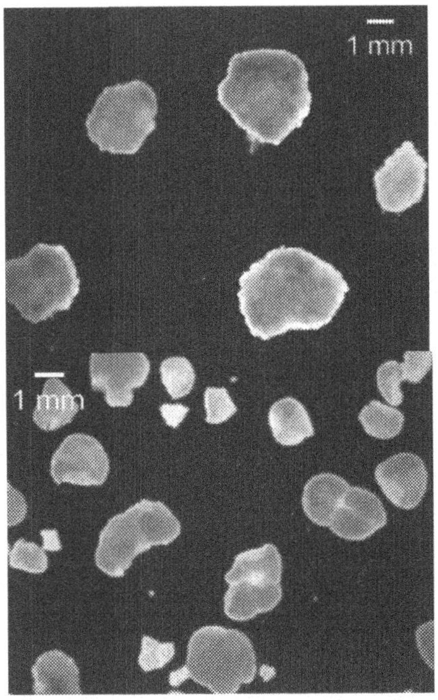

Figure 4 Granules grown with a short feeding time (a) and with a long feeding time (b).

During the aeration period of a cycle, oxygen will only partly penetrate into the granule, since it is consumed by the heterotrophic and autotrophic organisms in the outer layer of the granule. Inside the granule, oxygen is absent, but nitrite and nitrate, produced by autotrophic organisms in the outer layer, can diffuse into the core of the granules. In this zone PHB is available for the denitrifying growth of the heterotrophic organisms present in this part of the granule. This mechanism of the oxygen limited granule, instead of the substrate limitation as exists in continuous fed (biofilm) systems, plays a crucial role in simultaneous COD and nitrogen removal as exists in the aerobic granule system. This oxygen-limited system is only achieved when the acetate is fed as a pulse and the concentration of acetate in the reactor is very high at the beginning of the aeration period.

The experiment of Beun, with a 3-minute pulse feed of acetate, was successfully repeated and the granules were stable for 270 days, after which the experiment was stopped. As in the experiments of Beun (Beun et al., 2002), COD was totally removed and nitrogen removal was 62%, with oxygen saturation during the famine period. Also the start-up of the reactor was repeated several times, leading to stable granules after 60 days. After inoculation with nitrifying organisms, the ammonia concentration in the effluent started to decrease.

3.2 Bottleneck of scale-up

The pulse feed is seen as an important prerequisite for granulation. However, at full-scale this 3 minutes feeding period would imply very large pumps. As an example, for a 120,000 p.e. (Population equivalent) STP (see Chapter 7), this would mean pumps with a maximum working capacity of 5.7 m^3/s, which only work for three minutes per hour in a three reactor system or to work with much more and smaller reactors. Also large storage tanks would be needed and since one of the main goals of this system is surface area reduction, the use of storage tanks has to be avoided. All these approaches would increase the investment and production costs as well as the complexity of the system.

To be able to use smaller pumps and to avoid the need for storage tanks, a system with three reactors and a three-hour cycle was designed. In this configuration, the feeding time should be one hour. If the first reactor is filled, the second and the third are in the other phases and at the end of the filling phase of reactor 1, the influent flow can be switched to reactor two, and so on. This one hour filling time would decrease the maximum pump capacity drastically, but could disturb the pulse feeding character.

To maintain the concept of pulse feeding, an anaerobic one hour filling period was proposed, followed by an aeration period of 112 minutes and a short settling and effluent withdrawal time (respectively 3 and 5 minutes). In this design, no conversions would take place during this first hour filling. When the aeration is switched on, the granules and influent would be mixed and the reactor would experience a pulse feed. In practice, phosphate-accumulating organisms (PAO) will become dominant in the granules, which convert acetate during the anaerobic period to cell internally stored polymers, also leading to a low specific growth rate and thus to stable granulation (chapter 3.3.2).

To investigate simultaneously the possibility of no effluent extraction point, but pressing the effluent out of the reactor with the incoming influent, the reactor was fed from the bottom. With the low flow rate, a plug-flow of influent was created through the bed, which pushed the effluent from the pores to the upper part of the reactor. A colour tracer in the influent showed that no mixing occurred between the treated wastewater and the new influent. In the lab-scale set-up, an effluent withdrawal from the reactor at 50 cm, was maintained, because of practical reasons (maintaining a closed system, off-gas circulation).

3.3 Results

The SBAR was operated for over 233 days, with a 60 minutes feeding period, in which the influent entered the reactor in a plug flow through the packed bed of granules. In this period oxygen was not controlled and reached full saturation during aeration. The first 65 days of the experiment were used for start-up of the reactor and to reach a steady state. The second period (day 87 – day 117) was used for cycle measurements with inhibited nitrification, which was done by dosing allylthiourea (ATU). The last period was used to obtain full nitrification in the reactor and to observe whether the granules maintained stable.

3.3.1 Granulation

The granulation process during start-up of the reactor was comparable to the process with a 3 minutes pulse feeding. After 4 days, the first granules appeared. These first granules had many filaments at the surface, but after 9 days, smooth granules started to appear. After 40 days, all granules were smooth, regular granules, although a few very big granules were observed (diameter 6 mm). From day 68, the granule size started to decrease again to a minimum average size of 1.1 mm (figure 4). Small particles, which have good settleability characteristics, are very advantageous for the system, because this means a large biofilm surface area and therefore a higher substrate uptake rate per litre reactor volume. Another advantage is the small biofilm depth and therefore a larger active part in the granule. However, the particle size cannot be controlled in the reactor yet and more research is needed to find out which aspects influence this granule size and more important how to control these. The average sludge volume index was 27 ml/g, which is lower than the SVI as was found in the experiments with the pulse feed of Beun (39 ml/g).

The density of the granules increased from 38 gram TSS per litre of biomass to 121 g/l at the end of the experiment. The total biomass concentration increased to 8.5 g VSS per litre, with an ash content of 41% (TSS = 14.3 g/l). The density and VSS concentration were similar to the experiments of Beun, but the ash content was much higher (17% in Beun's experiments). Due to the anaerobic feeding phase, phosphate-accumulating organisms become dominant in the system. The stored poly-phosphates result in these high ash contents in the granule. The SRT was similar to the previous experiments, fluctuating around 40 days in the steady states of both experiments.

3.3.2 Conversion Processes

The anaerobic feeding through the bed, combined with aerobic periods for growth and ammonium removal, are circumstances in which phosphate accumulating organisms (PAO) are favoured. PAO are capable of taking up easy degradable substrates as acetate, under anaerobic conditions and store them as polyhydroxyalkanoates (PHA), like PHB and to lesser extent PHV (polyhydroxyvalarate). The energy of this transport and storage is supplied by the hydrolysis of intracellular stored poly-P to orthophosphate, which is released from the cell into the liquid. The reduction equivalents required for the conversion of acetate to PHB are supplied by conversion of glycogen, which is stored during the aerobic period from PHB (Mino et al., 1995, Smolders, 1995, Brdjanovic, 1998). In the aerobic or anoxic phase, without the presence of an external substrate, PHA is used as substrate for cell growth, polyphosphate synthesis and glycogen formation. The anaerobic and aerobic metabolisms are schematically shown in figure 5. Since the acetate is fed through the packed bed, the PAO are capable of anaerobically consuming the acetate and using their storage products during the aerobic period. Besides PAO, glycogen-accumulating organisms (GAO) can be present in an anaerobic/aerobic system, because of their capability of converting acetate during the anaerobic period without releasing phosphate. When the aeration is started, no acetate is left for the heterotrophic organisms. Therefore, only organisms capable of anaerobic acetate storage can grow in this system.

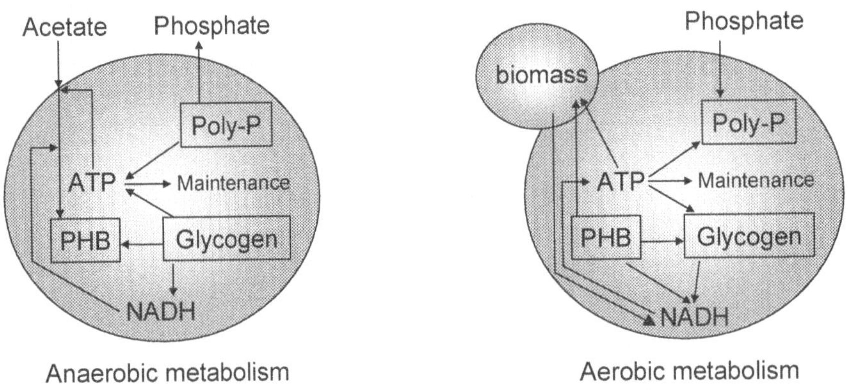

Figure 5 Metabolism of the phosphate accumulating organisms (Smolders et al., 1994a).

After 53 days of operation, the first decrease in effluent phosphate concentration was observed and after 68 days, the phosphate in the effluent was around zero (Figure 6). Phosphate stored as poly-P was removed with biomass present in the effluent. The biomass growth and thus the P-removal should be sufficient to take up the extra influent phosphate.

Also all dosed acetate was consumed after the anaerobic period. The phosphate release per acetate taken up was 0.86 g P/g C. Smolders reports a ratio of 0.67 to 1.96 g P/g C, depending on the pH (pH ranging from 5.5 to 8.5) (Smolders et al., 1994b). These ratios were measured with batch tests with sludge from a SBR in which only PAO were present. The value measured in the aerobic granular sludge was within this range, but concerning the pH range the ratio is low. It is most likely that other organisms were present, like GAO. The average growth yield of the biomass was 0.16 g VSS/g COD (or 0.27 g SS/g COD), which was calculated from the biomass increase divided by the acetate consumption (the effluent VSS concentration, plus the increase of the biomass concentration in the reactor). This value is much lower than the observed growth yield during operation with a short feeding time (0.3 g VSS/g COD or 0.39 g SS/g COD). In literature the same yield values for PAO are reported, for example 0.23-0.27 g TSS/g COD for a denitrifying Bio-P biofilm system (Falkentoft, 2000) and 0.25-0.3 g SS/g COD for a denitrifying bio-P system and 0.35-0.4 for an aerobic bio-P system (Kuba et al., 1993). The low growth yield at the SRT of 40 days was sufficient for the complete removal of 20 mg P/l from the influent. Biomass wash-out form the reactor and thus phosphate removal partly occurs via fines, eroded and sloughed of the granules by shear forces and particle-particle collisions. It is important that the phosphate content of these fines is high enough to remove the surplus phosphate. The effluent concentrations of phosphate in solution showed that these fines contained enough phosphate to remain sufficient phosphate removal.

Nitrification was inhibited with ATU during the first 100 days in order to allow a stable bio-P build-up. After the ATU dosing was stopped it took 39 days to reach full nitrification (no nitrite in the effluent). The high nitrate concentrations in the effluent were due to insufficient total nitrogen removal (average 24%) at these high oxygen concentrations (chapter 4). Also no denitrification occurred during the feeding time, since the plug-flow feeding pattern, so hardly any mixing between COD from the influent and nitrate still present in the reactor occurred.

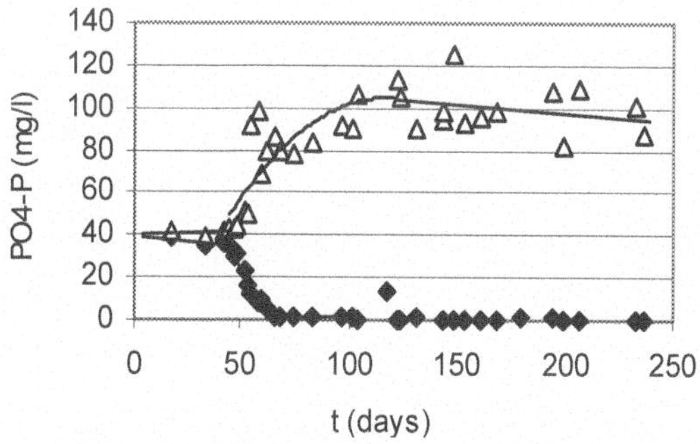

Figure 6 The phosphate concentration in the effluent (◆) and after the feeding phase, t=60 (Δ)

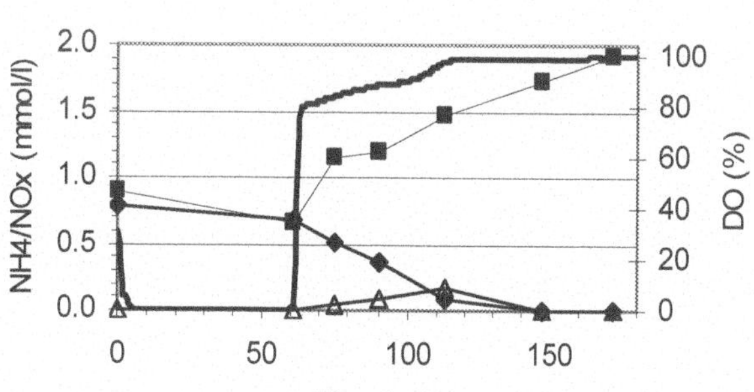

Figure 7 A cycle measurement of the airlift reactor at day 199 with 60 minutes feeding.
DO (-); NH_4 (◆), NO_2^- (Δ), NO_x (■),

3.4 Conclusions

Operation of the reactor with a long feeding time through the settled bed of granules appeared feasible and is favourable over a short feeding time. The two most important reasons are:
The application of a long feeding phase at large scale treatment plant, will simplify the batch scheduling, the influent dosing and will make a storage tank not necessary.
Due to the alternate aerobic and anaerobic phases, PAO will outcompete fast growing heterotrophic organisms, leading to stable granules and full phosphate removal.

It can be said that the long anaerobic feeding period, most favourable with a plug-flow regime, is very promising for the future application of the system at large(r) scale.

4 INFLUENCE OF THE DISSOLVED OXYGEN CONCENTRATION

4.1 Introduction

The distribution of biomass inside the biofilm or granule affects the performance of the processes taking place in the reactors. In aerobic biofilm systems, where COD removal and nitrification take place simultaneously, heterotrophic and autotrophic micro-organisms compete inside the biofilms for space and oxygen. When cell-external substrate (acetate) is present during the aerobic period, fast growing micro-organisms (heterotrophs) will be located in the outer layers while slow growing micro-organisms (autotrophs) will be confined to deeper zones of the biofilm with lower oxygen availability (Van Loosdrecht et al., 1995; Okabe et al., 1996). In the core of the granule, the oxygen concentration will be zero, but the nitrate produced by the autotrophic organisms will be available as electron acceptor for the heterotrophic organisms located in the deeper layers (figure 1, chapter 1).
In a system with an anaerobic feeding period and an aerobic reaction period phosphate accumulating organisms (PAO) become dominant (chapter 3). Contrary to the first mentioned granule system, in such system the aerobic outer layer can consist of a mixed system of PAO and autotrophs, because of their similar growth rates (respectively 1 – 1.5 d^{-1} and 0.8 - 1 d^{-1} (Brdjanovic et al., 1998, Henze et al., 1999). PAO can be divided into two different categories (Meinhold et al., 1998):

- Strict aerobic, unable to utilise nitrogen (nitrate or nitrite) as electron acceptor;
- Facultative aerobic (able to use oxygen or nitrogen as electron acceptor);

The first group can only be present in the outer layer of these phosphate-accumulating granules, the last group can also be present in the deeper layers.

Granular sludge growing at a high oxygen concentration shows a good COD and ammonium removal, but the nitrogen (nitrate) removal is insufficient (average of 24%). The type of feeding (long or short anaerobic feeding) has no influence on this removal capacity. In Dutch practice, larger STPs need to comply with an effluent standard of 10 mg total N/l (as an annual average). These standards might become stricter in the future, requesting even higher N-removal efficiency. With an average influent total Kjeldahl Nitrogen (TKN) of 55 mg/l, a total N-removal efficiency of 81% is needed. This can be partly done by a pre- or post treatment, but a large part should be removed by biological processes (>95% of the 45.7 mg/l dissolved N-kj) (chapter 7).

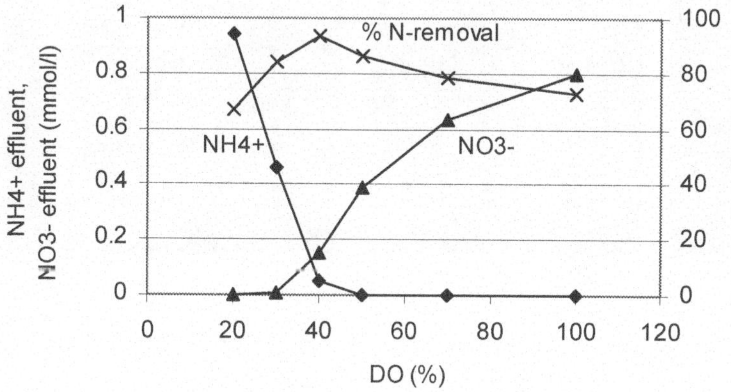

Figure 8 NO_3^- and NH_4^+ concentration in the effluent and percentage of N-removal as a function of the DO level in one cycle as calculated by Beun

An increase in nitrogen removal in a granule system can be achieved by lowering the dissolved oxygen concentration in the reactor. Preliminary modelling of the system with a pulse feed (no PAO present), showed that a dissolved oxygen concentration of 40% of air saturation would lead to a maximum nitrogen removal. Higher oxygen concentrations might lead to nitrate in the effluent, lower concentrations to ammonia (figure 8, Beun et al., 2001).

For practical and economical reasons, a lower oxygen concentration is needed to reduce energy costs. It is however unclear whether a low DO concentration has negative effects on granule formation. Therefore, the long-term DO effect on granule formation and stability was investigated. Both modes of feeding patterns were included in the study.

4.2 Low DO and pulse feed

4.2.1 Granulation

A reactor was operated during 143 days (Stage I) under the same operational conditions as described by Beun *et al.*, 2000a, except for the feed composition (lower COD/N ratio; 7.9 instead of 14.3 as used in the experiments of Beun). The results (conversions and granulation) obtained during this first period were similar as the results Beun obtained and are described in chapter 3.

After 143 days the DO concentration was reduced to 40% of oxygen saturation (Stage II) in order to obtain higher nitrogen removal. After ten days, growth of filamentous bacteria was observed on the surface of the granules. From day 174 the granules started to disintegrate, granule size and shape factor decreased and biomass loss was observed (Figure 9a). The SVI was around 48-75 ml/g TSS (Stage I) reaching a maximum value of 100 ml/g TSS at the end of Stage II. Although for flocculent sludge this might be good, under the conditions in the granular sludge reactor the sludge is slowly washed out. This change in the granule structure can be attributed to limitation of oxygen in the inner parts of the granule, where it was previously available operating with DO 100%. These observations are in line with results reported by Picioreanu (1999). He reported that stronger substrate gradients (like when the DO concentration is reduced) lead to less compact biofilms. The biomass washout caused a decrease of biomass concentration in the reactor from 5 to 3 g VSS/L. At the end of the experiment, the sludge retention time (SRT) decreased sharply from 25 days to 8 days. The experiment was stopped at this point.

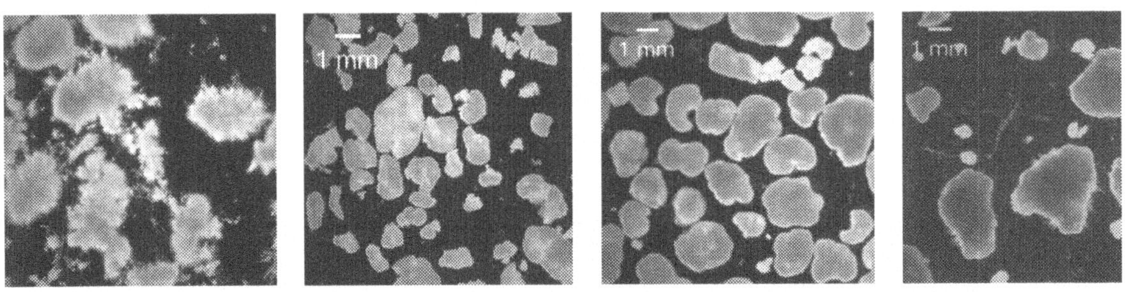

Figure 9 Granules at different DO concentrations; Short feeding period and DO 40% (a); long feeding period with DO 100% (b), DO 40% (c) and DO 20% (d)

In a second experiment, the reactor was started up again with a 3-minute pulse feed, but with a DO concentration of 40% of the saturation value (stage III). The maximum biomass concentration reached was 0.9 g VSS/l, so the COD load per gram of biomass was much higher compared to former experiments. According to previous studies increasing the loading rate leads to weaker and less dense biofilms (Picioreanu, 1999, Kwok et al., 1998). Combined with a stronger oxygen gradient, this apparently prevents the formation of granular sludge. The SVI was very high (200 ml/g TSS) and the density of the particles was only 13 g $TSS/L_{granules}$. After 1 month the reactor was stopped due to the apparent impossibility of formation of stable granules under these conditions.

4.2.2 Conversion processes

At high DO, ammonium removal was 100%. However, the total N-removal was only 8%, presumably due to biomass formation. After the reduction of the DO to 40% (Stage II), first the denitrification efficiency increased and NO_x^- was almost completely depleted, while the nitrification efficiency decreased and ammonia was present in the effluent. At the end of this Stage II, the ammonia was again completely nitrified, but nitrate and nitrite were not fully denitrified, although the N-removal capacity was higher (37%) than at DO 100%.

4.3 Low DO and long anaerobic feeding

The experiment with a long feeding period, showed a population shift towards phosphate accumulating organisms (PAO) (chapter 3). Since these organisms store substrate during the anaerobic period, there is no influence of oxygen on the conversion from easily biodegradable substrate to slowly biodegradable PHB, as in the system with a pulse feed under aerobic conditions. Therefore, the influence of DO on granule stability is expected to be of less importance. The experiments as described in chapter 4.2, were repeated with a long feeding period.
The airlift reactor with a long feeding time has been operated as a stable system for 234 days at a DO concentration of 100% (chapter 3), for 184 days at 40% and for 207 days at 20%.

4.3.1 Granulation

A decreased oxygen concentration in a system dominated by PAO's had only small effects on the granule characteristics. These characteristics are given in table 3 together with the results of the pulse feed experiment.

The average dry weight in the reactor increased with lower DO concentrations; 12 g VSS/l at DO 40% and 16 g VSS/l at DO 20%. The ash content of the particles was stable and around 34% (a sludge content of 25 g TSS/l). The increase in dry weight in the reactor showed a dependency with the SVI and the granule diameter. In time, the settled bed filled up to the effluent withdrawal. In the present experimental set-up, the excess sludge produced was washed out with the effluent. The lower the particle diameter, the more particles fit in the reactor in the settled bed (packing of the particles). This causes also a decreased SVI and the lower the SVI (18 ml/g at DO 20%), the less biomass washed out. This led to an increase in total dry weight in the reactor.

The granule structure (size and shape) was changing in time and could not be controlled as such. However, the granules were stable without fluffy and filamentous structures, as observed with the short feeding time (figure 9). In general, it can be concluded that to be able to apply lower oxygen concentrations in this aerobic granule system, selection for slow growing organisms has to take place, which can be controlled by the feeding regime. Only then the granule structure will be dense and smooth. The reasons for the change in particle size distribution and particle shape of dense and smooth particles have to be investigated in later research.

4.3.2 Conversion processes

Table 3 illustrates the measured nitrogen removal at different DO concentrations. At lower DO concentrations, the nitrate concentration in the effluent decreases and the ammonia concentration in the effluent increases slightly, but remains low in all cases. After each change in DO concentration, it took approximately one sludge-age to react on the changed situation. Fist ammonia started to appear in the effluent, which was replaced by nitrate when the oxygen concentration remained low and population in the granule started to shift. The lowest concentration (D=20%) showed the same pattern, with the main difference that the nitrate concentration in the effluent remained low and the total N-removal in the system was 90% on average over the last 100 days of operation (Figure 10). The model of Beun predicted an increased amount of ammonia in the effluent at DO 20% and an optimal N-removal capacity at DO=40%. This model was based on a 3 minutes pulse feed and thus on a clear heterotrophic outer-layer of the granule with high growth rates. Since the heterotrophic PAO and the autotrophic nitrifyers have a comparable growth rate, these organisms are likely to be mixed in the outer layers and thus oxygen is more readily available for nitrification at DO 20%. The inner part of the particle uses enough nitrate for the conversion of PHB to maintain low total-N concentrations in the effluent.

The SRT of the granules was 37 days at DO 100%, 64 days at DO 40% and at DO 20% this value increased to 145 days, after which it decreased to 87 days. Because of this very high SRT, not enough phosphate was washed out with the effluent biomass and a 96% P-removal could not be maintained. With an increasing sludge age, the P-removal decreased to 40%. As soon as the SRT decreased again, the P-removal increased to 94%. This shows that in order to obtain a long-term sufficient phosphate removal, the sludge age has to be controlled by removing appropriate quantities of granules from the system.

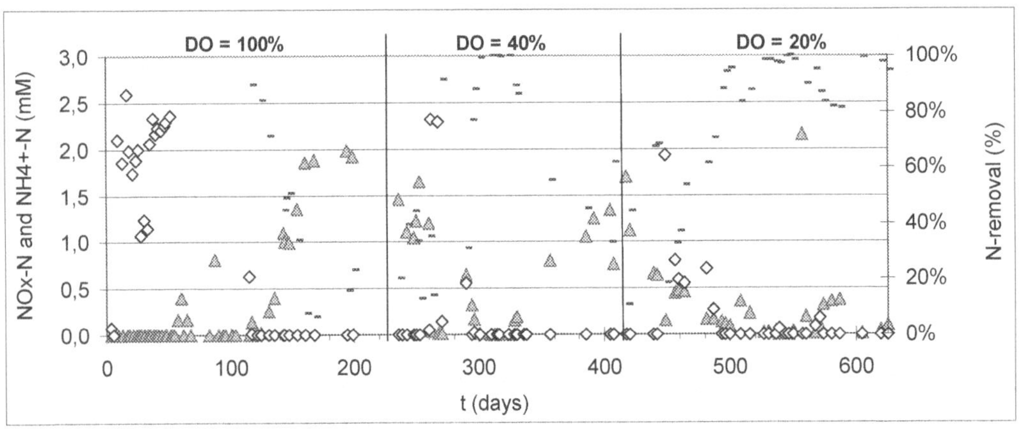

Figure 10 The effluent NO_x (▲) and NH_4^+ (◊) at different dissolved oxygen concentrations and the total nitrogen removal in percentage of the influent ammonia (-)

Table 3 Results of the experiments with different oxygen concentrations and feeding phases (3 and 60 minutes)

Averaged values	Short feeding period			Long feeding period		
	DO 100% (I)	DO 40% (II)	DO 40% (III)[2]	DO 100%	DO 40%	DO 20%
Feast phase (min)	15.5	20	160	--	--	--
Bed volume (l)	0.8	0.9	0.6	1.1	1.1	1.3
SRT (days)	25	8	ND	37	60	114
Density granules (gTSS/lgranule)	53	60	12.6	95	71	78
Equivalent diameter (mm)	1.56	ND	5.0	1.28	1.12	1.05
gVSS/l (reactor)	5.1	3	0.9	8.5	12	16
GTSS/l (effluent)	0.07	0.10	0.14	0.07	0.05	0.05
SVI (ml/gTSS)[1]	50	100	200	24	20	18
mg TOC/l (influent)	105	105	105	105	105	105
mg TOC/l (effluent)[1]	20	20	ND	27	12	20
mg NH_4^+-N/l (influent)	48	48	48	48	48	48
mg NH_4^+-N/l (effluent)	0	1	ND	0	0	0.1
mg NO_3^--N/l (effluent)	35	21	ND	24.8	17	3
mg NO_2^--N/l (effluent)	0	1.5	ND	0	0	0
mg PO_4-P/l (influent)	--	--	--	20	20	20
mg PO_4-P/l (effluent)	--	--	--	0.56	0.55	3.9

ND Not detected.

[1] *after 5 min of settling.*

[2] *Values taken from the period when only granules where present but very variable due to wall growth.*

4.4 Conclusions

Operation of the reactor at a low dissolved oxygen concentration is only feasible when a selection of slow growing organisms takes place. This means that a long anaerobic feeding time should be applied, in order to favour the growth of slow growing PAO over the growth of fast growing heterotrophic organisms. With a long anaerobic feeding period, stable granules will be formed.

Nitrogen removal increases with decreasing oxygen concentrations. At the lowest tested DO (20%), N-removal was sufficient to reach the effluent standard (total N<10 mg/l). In the effluent, only nitrate was present and ammonia could not be detected. This means in practice, even lower oxygen concentrations might be feasible. The possibility of applying low oxygen concentrations will reduce aeration and power costs.

Phosphate removal should be maintained by controlling the sludge age or solid retention time. If the SRT becomes too high, phosphate will not be properly removed.

It can be said that a low oxygen concentration combined with a long anaerobic feeding period is very promising for the future application of aerobic granular sludge at large(r) scale.

5 INFLUENCE OF THE REACTOR CONFIGURATION

5.1 Introduction

The experiments with aerobic granulation were performed so far in sequencing batch airlift reactors (SBAR). The disadvantage of this reactor type is the complicated and more expensive construction on large scale, when it is compared to a bubble column or aeration tank. Beun (1999) showed that in a sequencing batch bubble column (SBBC) fed with a three-minute pulse feed, less dense biofilm structures (ρ = 12 g TSS per litre of granules) were obtained. In an SBAR biofilm densities of 60 to 120 g X/l biomass were obtained. Stratification of granules in the bubble column caused a lower shear, originating from particle-particle interactions, in the upper part of the column. The fast growing organisms in the outer layer of the granule have the chance to form filaments and, therefore, less dense biofilms. This effect was intensified, because these less dense particles remained in this upper part of the reactor and were exposed to decreasing shear stress (Beun et al., 1999). In an airlift reactor, this effect does not take place, because of the excellent mixing of the particles in these type of reactors and the higher shear forces on the granules. Since the long feeding phase the selection for slow growing phosphate accumulating bacteria, the need for shear becomes less evident.

In order to investigate the necessity of an airlift reactor (the need of a riser) to generate stable and dense granules, a SBBC was started up. All process variables were kept the same as in the former experiments (chapter 3 and 4), only the riser was removed from one of the reactors.

5.2 Results

5.2.1 Granulation

A SBBC was started up with a dissolved oxygen concentration of 40%. The granulation process in the SBBC was slower than in the SBAR. A week after start up in a SBAR, the first granules are observed and after 2 weeks they have an average diameter of 1 mm. The granules stabilise after 40 days at 1.3 mm (average diameter fluctuating between 1.1 mm and 1.6 mm at DO 100% and a maximum diameter of 1.2 mm at DO 40%).

In the SBBC, the first month the granules tend to form flocs during settling. After 27 days it was observed for the first time that the biomass only consisted of very small separate granules (average equivalent diameter was 0.433 mm). After 2 months, the particles reached an average diameter of 1 mm. The granules increased in diameter to an average of 1.3 mm after 3 months (fluctuating between 1 and 1.3 mm). The shape factor[1] of the granules varied between 0.67 and 0.74 and the aspect ratio[2] between 0.70 and 0.76, which is comparable to the values obtained with the SBAR (respectively 0.59 to 0.73 and 0.67 to 0.73).

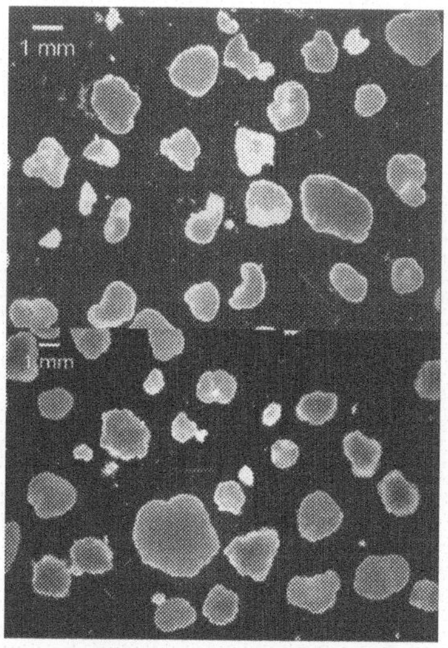

Figure 11 Stable granules in a SBAR (a) and in a SBBC (b).

[1] 0 = very irregular structure; 1= smooth and round structure
[2] 0 = line; 1 = round

Table 4 shows the summary of results of the granules described above. All results show great similarity to the results of other researchers in their biofilm studies and what is stated in theory (Picioreanu, 1999, Tijhuis et al., 1995). By applying a long anaerobic feeding period to the granules, selection for slow growing phosphate accumulating bacteria takes place, resulting in dense, smooth and thus well settling granular structures. Therefore, the strong shear stress of the airlift is not necessary when the long feeding period is applied.

Table 4 Differences in granule characteristics between the two reactor types, feeding patterns and dissolved oxygen concentrations

Particle characteristics	pulse feed			long feeding period		
	SBAR		SBBC[1]	SBAR		SBBC
	DO 40%	DO 100%	DO 100%	DO 40%	DO 100%	DO 40%
Stability of granules	not stable	stable	n.a.	stable	stable	stable
Shear in reactor	High	high	low	high	high	low
Average growth rate	High	high	high	low	low	low
Average diameter (mm)	5.0	1.56	2.0	1.12	1.28	1.14
Biomass density in the particle (g TSS/ l biomass)	13	53	12	71	95	101
Dry weight in reactor (g VSS/l reactor)	0.9	5.1	3	12	8.5	13
SVI (ml/g)	200	50	n.a.	20	24	19

[1] (Beun, 1999b)

5.2.2 Conversion processes

The application of a SBBC instead of a SBAR has no influence on the conversion processes of the granules. The acetate is totally converted after the anaerobic feeding period. The average phosphate removal is 93%. The N-removal was not very stable because of problems with the oxygen control, but reached values of 95% and was on average 83% during the stable periods. Because of the DO concentration of 40%, the nitrogen was present as NO_3^- in the effluent.

5.3 Conclusion

The only difference in granulation between a bubble column and an airlift reactor was found in the duration of the start-up period, granule characteristics were highly comparable. As long as a slower start up of the reactor is not a problem, the use of a SBBC, combined with a long feeding period is feasible and has no negative influence on the granulation. Extra attention has to be given to the mixing of the granules and to the stability of the conversion processes.

6 EXPERIMENT WITH PRE-TREATED SEWAGE

6.1 Introduction

The past results of aerobic granular SBAR and SBBC, as described in the former chapters, have been very promising for the use of this type of reactor in practice. Simultaneous full COD removal, high phosphate removal and high nitrogen removal have been reported. Also size and shape of the granules changed in time, but these characteristics were stable at different oxygen concentrations, when a long anaerobic feeding period was applied. The tests have been carried out with the use of synthetic influent, consisting of an acetate/ammonia mixture. No complex substrate and no suspended solids were present in the influent.

The next step of the design of a full-scale reactor is a pilot-phase, but before the investments of a pilot scale can be made, a test with presettled wastewater has to be performed to investigate the formation of granules on more complex influent. The behaviour in the reactor of suspended solids from the sewage is not investigated at this scale, since wall effects and shear stress will be too different from a full-scale installation.

6.2 Selection of the influent

6.2.1 Pre-treatment

For the experiment with municipal wastewater, an influent was used that is very close to an actual pilot or full-scale situation, since this will give the best insight in the feasibility of the system. Wastewater was collected from the STP, right after the screen (5*25 litre per batch). The tanks were stored in a 4°C room, where the settlable suspended solids settled in the tanks. Once per two days, 25 litres of this influent was restored in a cooled storage tank (10°C) near the reactor. Per cycle 1.5 litres were pumped into a 2-litre storage tank to warm up to room temperature and subsequently dosed to the reactor. In this way a good pre-settled influent is prepared, without the use of flocculating agents. Pre-treatment of the raw wastewater was chosen, to avoid clogging of the tubes and other reactor parts. The use of flocculating agents was avoided, since their influence on granulation is not known and their use might lead to different results. In this experiment this was the step from relatively simple to more complex influent.

6.2.2 Selection of the raw sewage

During the first period (day 0 – 28) of the experiment, wastewater from STP Gouda (pipeline of Stolwijk) was used. The influent concentrations analysed during the fist period of this experiment showed a much more diluted wastewater than was expected.
A new influent with higher COD values was taken from the treatment plant of Berkel. This influent is applied from day 29 of the second start up. The total biodegradable COD load is still 31% lower than with the synthetic influent.
Table 5 shows the different used influent concentrations, compared with the synthetic influent, used in former experiments and the wastewater characteristics, used for the design of the large-scale reactor. The influent concentrations of the pre-treated wastewater, differ a lot from the synthetic influent. Especially the biodegradable COD concentrations are much lower (Table 5), resulting in a slow increase in biomass concentration. Potentially, always some intrinsic biomass washout occurs. When the COD load is low, a relatively high washout occurs in terms of washed-out biomass COD per influent COD. Therefore a longer time is needed before a clear

increase in biomass will be observed, which results in a longer period needed for starting up the system.

Table 5 Influent characteristics

	Synthetic influent, first lab tests	Standard raw influent (STOWA)	Average presettled influent concentrations of	
			treatment plant Gouda [2]	Treatment plant Berkel[3]
Suspended solids (mg/l)	0	250	84	75
BOD_{∞} (mg/l)	-	220	-	70
soluble COD (mg/l)		216	186	253
COD total (mg/l)[1]	381	600	223	308
VFA's (mg COD/l)		-	44	85
PO_4^--P (mg/l)	19.6	9	5.35	8.5
NH_4^+-N (mg/l)		-	31.6	57.1
Total N, filtered (mg/l)	50	45	42.2	67.2
Total N, unfiltered (mg/l)		55	-	71.1
Total COD/Total N	7.6	10.9	5.3	4.3
Total COD/Total P	19.4	66.6	41.7	36.2

1) In case of the lab experiments, this is the value after settling in the 4°C room.

2) The average values of Gouda (Stolwijk pipeline) are based on 2 batches.

3) The average values of Berkel are based on 8 batches.

6.3 Results

6.3.1 Granulation

The reactor was started up with wastewater from the treatment plant Gouda (table 5, table 6). The applied settling time of 5 minutes was too short, since the flocs (with granules) coagulated when aeration was stopped, and the settling velocity of the biomass was therefore too low to maintain the biomass in the system. Therefore, the reactor was started up again after 13 days of operation (=day 0) with new activated sludge. The reactor was also frequently inoculated with the dispersed cells and the little flocs from the effluent from the stable granular SBAR 2, operated with synthetic wastewater in order to enrich the biomass with PAO and autotrophic organisms.

The settling time was varied between 15 and 6 minutes, in order to keep granules with settling rate between 2 m/h and 5 m/h in the reactor. This was done in order to prevent too much biomass loss by wash out with the effluent. After 10 days, a riser was added to the SBBC, in order to speed up the granulation process (chapter 5).

Table 6 Operation of reactor 3, changes in settling time, influent and other changes in operation

Day	Settling time	Wastewater	Other Changes
-13	5 minutes	Gouda	start up first time
0	15 minutes	Gouda	start up second time
10	7 minutes	Gouda	from bubble column to airlift
6-30	varied between 6 and 7 minutes	Gouda	
30	15 minutes	Berkel	
50	10 minutes	Berkel	Total cycle time shortened to two hours to increase COD load
69	6 minutes	Berkel	

The increase in biomass content in the reactor with a low COD load is very difficult, since an intrinsic biomass wash-out will always occur (chapter 6.2.2). In practice at a wastewater treatment plant, the total cycle time would be shortened in order to increase the COD-load on the system. This method was not feasible at lab-scale, since the amount of available wastewater was limited. Therefore, the influent was changed to a more concentrated variant (Berkel) after 1 month of operation. Short after the change to the more concentrated wastewater, the granules did not flocculate anymore during settling and the granules started to increase in diameter to 0.5 to more than 1 mm after 40 days (figure 12). This is the first proof that granulation can occur with pre-treated municipal wastewater.

The development of the granules was slower than in the reactors operated with synthetic wastewater, mainly due to the lower availability of (easily biodegradable) COD. This improved when the cycle-time was decreased to 2 hours with the same amount of influent per cycle as in the 3 hour cycle (1600 ml/cycle), which caused an increased COD load from 1.16 to 1.74 kg/m^3/day (after storage the COD concentration in the wastewater decreases, leading to a COD load of 1.28 kg/m^3/day after 5 days of storage). The size of the granules reached an average equivalent diameter of 1.1 mm after 70 days, the aspect ratio 0.69 was and the shape factor 0.44, which is lower than on synthetic influent (respectively 0.70 and 0.73). The average SVI was 38 ml/g. The experiment was stopped while the biomass concentration in the reactor was increasing rapidly, but the last measured biomass concentration was 2.2 g TSS/l at day 100. Figure 13 shows a particle size distribution curve (day 63). Contrary to the smooth particles grown on synthetic wastewater, the surface of many granules grown on pre-treated wastewater were covered with small filamentous structures (figure 12).

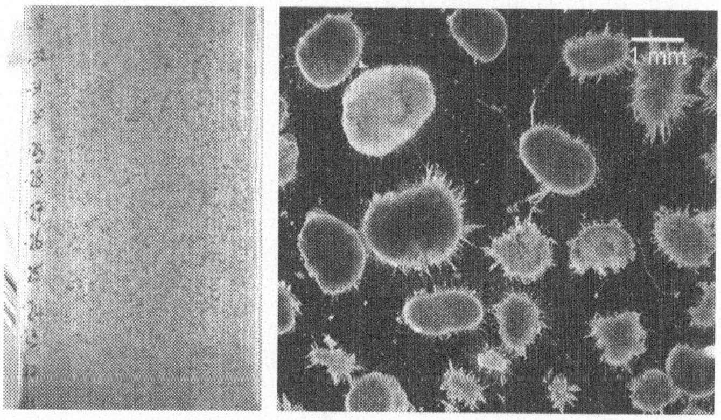

Figure 12 Settling granules in the reactor at day 42 (a). Granules after 63 days (b) of operation

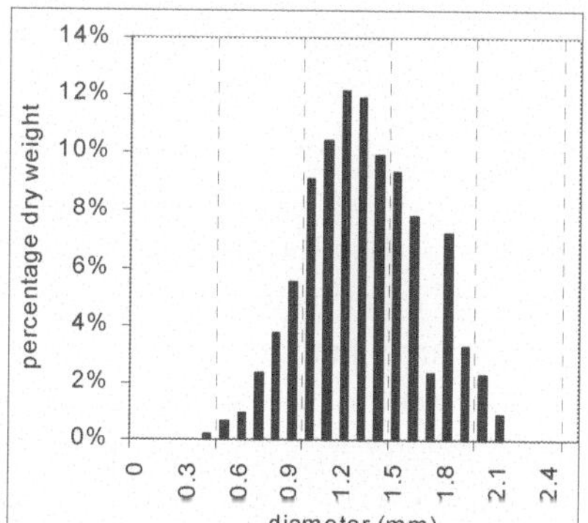

Figure 13 Particle size distribution of the granules at day 63

An explanation for the growth of filaments on the surface of the granules can be the presence of COD during the aerobic period. The fast degradable substrates (as acetate) are taken up during the anaerobic period, but much of the complex substrate and the low amounts of particulate COD are still available during the aerobic period. This favours fast growing heterotrophic organisms and among these filamentous organisms, that can use this substrate to grow during the aerobic period.

Shear can play an important role in the appearance of an irregular surface as well. Since only very small amounts of biomass are present in the reactor, there are very few particle/particle collisions. These particle/particle collisions play an important role in the amount of shear that the granules experience (Gjaltema et al., 1997). In previous research it was observed that filamentous growth on the surface of the granules increase with decreasing shear stress (Beun et al., 1999). In that case, the problem of filaments at the surface of the granules will be minimised when more granules are present in the system and thus more particle/particle collisions occur.

The amount of wall-growth was a problem during the experiment. Since an airlift reactor was used, a surface area of 0.5 m^2 was available for wall-growth (167 m^2/m^3). The biofilm quantity grown on the surface area of the reactor (measured once) plus the biomass in the effluent and the increasing dry weight concentration in the reactor, approximates the theoretical growth yield of 0.3 g SS/g COD for PAO (Kuba et al., 1993). Because of the extensive wall growth and the wash-out of organisms, the settled bed only increases with 0.35 cm per day, which is lower than as observed in the experiments with a synthetic wastewater. This extensive wall-growth is a typical small-scale problem. Such slow increase in biomass concentration in the reactor is not to be expected in a large- or pilot-scale system.

6.3.2 Conversion processes

The operation time of this experiment was too short to get stable conversion processes. The conversion processes derived from the reactor during operation at a cycle time of 2 hours and the wastewater from RWZI Berkel, will be given in this paragraph.

The VFA are totally removed during the anaerobic feeding period. The COD concentrations in the effluent were on average 0.17 g/l. This means that 60% of the biodegradable COD was converted. This biodegraded COD consists partly of particulate and colloidal material. At pilot-scale, this has to be improved and further investigated.

The P uptake per gram of biomass fluctuated largely, but during the last days of operation an average value 0.8 mg P/g d.w. was measured. In phosphate removing Bio-P systems values in the range of the 5-7 mg P/g d.w. reported by STOWA (STOWA, 2001). The sludge wash-out and wall-growth can have negative effects on the development of the slow growing PAO, as well as the growth of heterotrophs during the aeration phase (chapter 6.3.1).

The total N removal fluctuated between 5% and 24%. Balances, made from the measurements, showed that nitrification and denitrification occur simultaneously. Also the average N-conversion per gram dry weight was 11 mg N/g TSS and even 26 mg N/g TSS at day 54. Compared to the airlift reactor on synthetic wastewater this is very high, since this reactor removes up to 3.8 mg N/ g TSS. It has to be mentioned that this ratio cannot be higher in this last reactor, because it is limited by the ammonia loading rate.

At the end of the experiment, the reactor was fed during one cycle with the synthetic wastewater as is used in the other experiments. This leaded to a full COD (acetate) removal during the anaerobic period, 51% N-removal (as ammonia in the effluent) and 40% P-removal. This result shows that part of the organisms are PAO and that the nitrate that is formed is immediately denitrified, since the amount of ammonia removed is too much to be only used for growth.

6.4 Conclusions

It can be said that the granulation process does occur with this complex influent. The granule formation was slower than on synthetic wastewater. This was due to the lower COD load of easily degradable substrate. Also, the granules were less regular and smooth. The formed granules are relatively small (1.1 mm) and have good settling characteristics (settling velocity > 5.5 m/h, SVI < 40 ml/g).

Biomass wash-out and wall-growth had a negative effect on the selection of the slow growing organisms and thus the processes of N-removal and P-removal. In a larger scale reactor, these effects will be less, since the ratio wall/volume ratio is much lower.
The COD, N and P loads have to be increased during the start-up, because this most certainly will have a positive effect on the growth of biomass.
From the results in the laboratory, we are not able to conclude that the conversion processes are sufficient for the treatment of the wastewater in practise. Results are promising though and the formation of granular aerobic sludge in this system is the main first step.

7 DESIGN AND ECONOMICS OF A FULL-SCALE GSBR TREATMENT PLANT

7.1 Introduction

Parallel to the research work a feasibility study was carried out for the full scale application of a GSBR treating municipal wastewater. For this purpose, two alternatives of the aerobic granular sludge technology are compared to three variants of conventional STPs based on activated sludge technology. Furthermore, costs determining process parameters are identified.

7.2 General basic conditions

7.2.1 Capacity and wastewater composition

For the capacity a range for the pollution load is taken between 60,000 and 190,000 population equivalents (136 g TOD). For the wastewater the standard composition defined in STOWA research report 98-29 is assumed.
The ratio of dissolved and suspended particles is based on the results of the STOWA research "Physical/chemical pre-treatment of wastewater" in which the colloidal parts (with a diameter between 0.1 and 5 µm) are added up to the suspended particles. A summary of the wastewater composition and flow rates is given in table 7.

Table 7 Wastewater composition and flow

Composition	dissolved (mg/l)	suspended (mg/l)	Total (mg/l)
COD	216	384	600
BOD	106	114	220
Suspended solids	n/a	250	250
TKN	45.7	9.3	55
P_{total}			9
Flow rate	**Unit**	**Value**	
Average flow	l/(pe.d)	160	
RWF	l/(pe.h)	34	
RWF/DWF ratio	-	2.5	

In table 8 the removal efficiencies of primary sedimentation with and without chemical dosing are summarised.

Table 8 Removal efficiencies of primary sedimentation with and without chemical dosing

Composition	Primary sedimentation	Primary sedimentation with pre-precipitation (Fe and polymer)	Primary sedimentation with polymer dosing
COD	31%	53%	53%
Suspended solids	51%	80%	80%
TKN*)	7%	19%	19%
P_{total}	7%	70%	38%
*) excluding the N-load in the rejection water of the sludge treatment			

7.2.2 Effluent requirements

The effluent requirements are based on the Discharge Decree for municipal wastewater in The Netherlands (see table 9). The requirements with regard to nitrogen and phosphate are determinative to a large extent for the technological design of STPs.

Table 9 Effluent requirements

Parameter	Value (mg/l)	Type of requirement
COD	125	daily average
BOD	20	daily average
Suspended solids	30	daily average
N_{total}	10	yearly average
P_{total}	1	moving average of ten successive samples

7.2.3 Flow pattern

Figure 14 shows the daily flow pattern of a specific Dutch STP with a capacity of 400,000 p.e. at dry weather conditions as well as at rain weather conditions. As appears from Figure 14 the maximum flow is based on a day with a 24-hour continuous maximum rain weather flow (RWF).

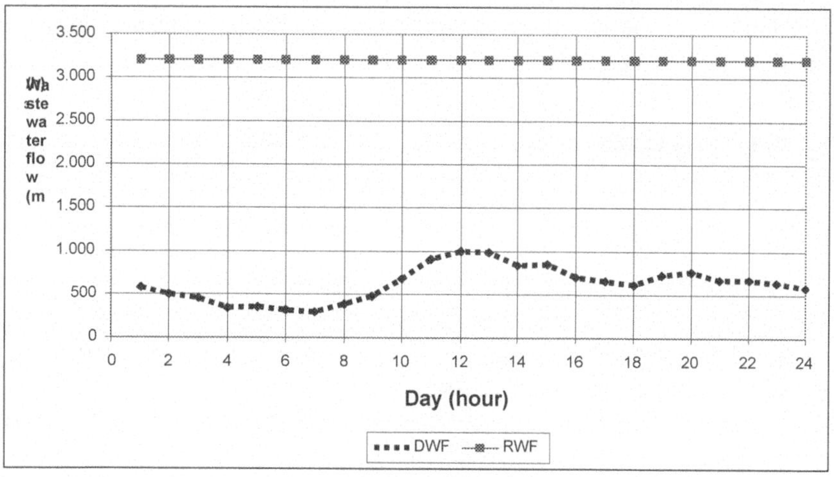

Figure 14 Flow pattern

7.2.4 Cost estimates

Table 10 summarises the starting points for the calculation of the capital costs. Capital costs are calculated based on annuities.

Table 10 Starting points calculation capital costs

Cost factor	Unit	Value
Depreciation period civil parts	year	30
Depreciation period mechanical parts	year	15
Depreciation period electro-technical parts	year	15
Interest rate	%	6

In table 11 the starting points for the calculation of the operational costs are given.

Table 11 Starting points calculation operational costs

Cost factor	Unit	Value
Electricity	€/kWh	0.054
Sludge disposal	€/tonDrySolids	320
Iron chloride (41%)	€/ton	115
Poly electrolyte (liquid, 50%)	€/kg active	3.6
Maintenance civil parts	% of investments	0.5
Maintenance mechanical parts	% of investments	2
Maintenance electro technical parts	% of investments	2

7.2.5 Footprint and land price

The footprint is calculated by the sum of the net surfaces of all process units and buildings, multiplied by a factor 1.3.
The land price is assumed to be 22.7 €/m^2. For the calculation of the annual costs associated with the purchase of land, only the interest costs are taken into account. Depreciation costs are not calculated, assuming that land keeps at least its original purchase value.

7.3 Reference systems

7.3.1 Introduction

The following three alternatives of the conventional activated sludge system (AS) are taken as a reference:

- a conventional AS-system with primary sedimentation, biological phosphate removal, additional simultaneous phosphate removal and sludge digestion;
- a conventional AS-system with primary sedimentation, pre-precipitation and sludge digestion;
- a conventional AS-sludge system with biological phosphate removal, additional simultaneous phosphate removal and simultaneous aerobic sludge stabilisation.

Global process flow diagrams of the aforementioned reference alternatives are presented in Appendix 1 (figure 1 to 3).

7.3.2 Design assumptions

The following design assumptions are made:

- design temperature : 10 °C;
- maximum hydraulic load primary sedimentation : 4 m/h
- Me/P ratio based on the influent P-load (mol/mol):
  - o AS with primary sedimentation, Bio-P and sludge digestion : 0.2
  - o AS with primary sedimentation, pre-precipitation and sludge digestion : 1.5
  - o AS with Bio-P : 0.2
- poly electrolyte dosage with pre-precipitation : 0.8 gPE_{active}/m^3;
- pre-denitrification volume : 20%
- Sludge Volume Index (SVI) : 150 ml/g
- hydraulic retention time sludge digestion : 20 d
- average degradation MLVSS : 47 %
- poly electrolyte dosage sludge dewatering : 12 $gPE_{active}/kgDS$
- Dry Solids content dewatered sludge : 25 %
- N-total removal calculations according to Kayser (1983) and STOWA (1997);
- sludge production calculation according to Chudoba (1985);
- aeration by means of surface aerators with an efficiency of 2.0 kgO_2/kWh in clean water;
- calculation of the oxygen demand according to Beuthe (1970);
- STORA-directives for the design of sedimentation tanks;
- gravitational thickening of primary sludge;
- mechanical thickening of surplus sludge;
- dewatering by centrifuges.

In table 12 the design sludge load of the aeration tanks are given for the reference alternatives. These design loads are based on the biological part of the sludge.

Table 12 Design sludge loads of aeration tanks in reference alternatives

Alternative	COD (kgCOD/(kgDS.d))	Nkj (kgNkj/(kgDS.d))
AS with primary sedimentation, Bio-P and sludge digestion	0.14	0.018
AS with primary sedimentation, pre-precipitation and sludge digestion	0.11	0.019
AS with Bio-P	0.16	0.014

7.3.3 Dimensions and costs

In appendices 3 to 6 the dimensions and cost estimates of the three reference alternatives for a capacity of 120,000 p.e are summarised. Operational costs are based on main cost factors such as sludge disposal costs, power use aerators, chemical use and maintenance costs. Power use of the aerators for the reference alternatives with sludge digestion is corrected for power production from biogas.

Figures 15 and 16 show the investment costs and the total specific annual costs (capital costs and operational costs). As expected, the investment costs of the reference alternative "AS with primary sedimentation, Bio-P and sludge digestion" are the highest (see figure 15). The reference alternative "AS with Bio-P" results in the lowest investment costs.

Based on the total specific annual costs the reference alternative "AS with primary sedimentation, pre-precipitation and sludge digestion" is the most costly alternative throughout the range (see figure 16). In comparison with the alternative "AS with primary sedimentation, Bio-P and sludge digestion" the operational costs are substantially higher. This is caused by a higher sludge production and a higher chemical use required for phosphate removal. Compared to the reference alternative "AS with Bio-P" the higher capital costs and higher chemical use cause the cost difference.

The reference alternatives "AS with primary sedimentation, Bio-P and sludge digestion" and "AS with Bio-P" are economically the most attractive and comparable. The reference alternative "AS with Bio-P" is the most economic alternative in the range between 60,000 and 150,000 p.e. In this range the low capital costs are decisive rather than the relatively high sludge production and high electricity use. The reference alternative "AS with primary sedimentation, Bio-P and sludge digestion" is the preferred alternative from 150,000 p.e. on. As from this capacity the lower sludge production and lower electricity consumption weigh heavier in the total costs.

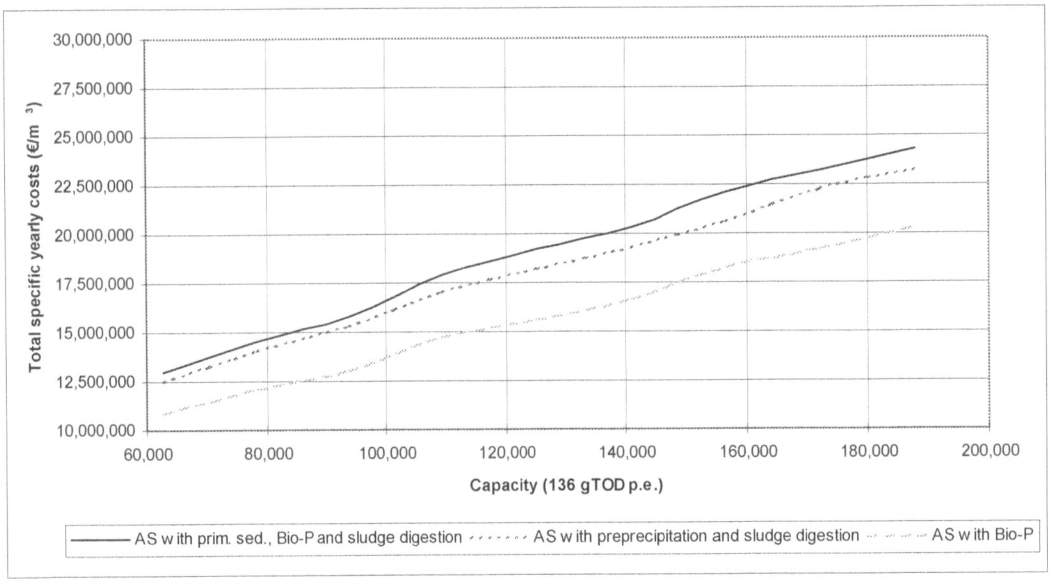

Figure 15 Investments reference alternatives

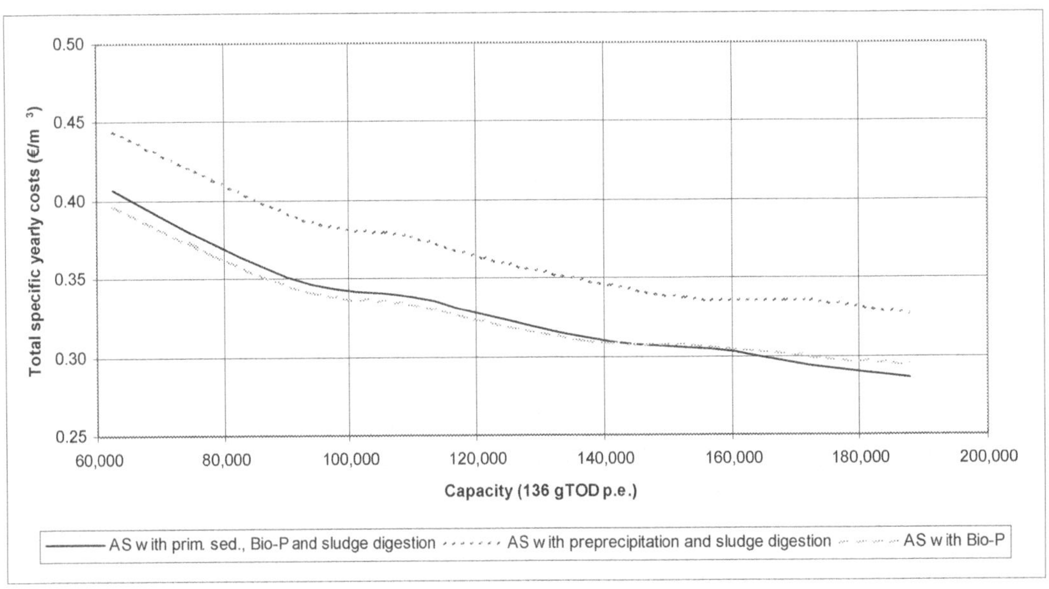

Figure 16 Total specific annual costs reference alternatives

7.4 Interaction between feasibility study and research

Because of the results from the feasibility study the research program was partly adjusted. Changes in reactor operation and their influences on the granulation are extensively discussed in the previous chapters. In this paragraph, the interactions between the laboratory research and the feasibility study are summarised.

Cycle configuration:

- *Filling time*: A pulse feed in three minutes as performed during the first experiments is not feasible for a full-scale application (the flow during filling of the GSBR would be a factor twelve higher compared to the RWF). Therefore, experiments with a longer feeding period were performed (chapter 3);
- *Sedimentation:* For a full scale GSBR with a construction height of 5-6 metres and a particle sedimentation velocity of 15 m/h, a sedimentation time of 3 minutes is too short and should be extended to 15-30 minutes. Short sedimentation times result in low and undesirable Height/Diameter-ratios of the reactor. No further investigations on this topic were performed;
- *Effluent discharge:* Short decantation times lead to high discharge flows and expensive discharge constructions, despite the fact that effluent is discharged by gravity. The dimensions of a post treatment step depend also strongly on the maximum discharge flow, although this can be compensated with buffer tanks. Research was carried out to test the feasibility of simultaneous feeding and effluent withdrawal: incoming feed would push out the effluent in the reactor, assuming a plug flow character through the settled bed of granules. The main conclusion of these tests is that further research has to be carried out (e.g. on pilot scale) before conclusions can be drawn.

Phosphate removal

Initially chemical phosphate removal was considered. This seemed logical in combination with extensive primary treatment with pre-precipitation. Design calculations showed higher costs for chemical phosphate removal in comparison with biological phosphate removal. This was due to higher chemical use and higher sludge production. Based on these calculations biological phosphate removal is preferred to chemical removal. It was demonstrated that a longer filling time leads to a microbiological population shift towards PAOs (see chapter 3) enabling biological phosphate removal.

Nitrogen removal and oxygen saturation

Experiments proved that DO concentrations of 20% of the saturation value result in high Ntotal conversions. In the presence of PAOs stable granule formation is possible under these conditions.

Sludge characteristics

Besides the granule characteristics (sedimentation velocities, size, density, etc.) the characteristics of the biomass which is washed out can also be important in case these 'fines' determine a large part of the sludge production. Important aspects in terms of sludge treatment are the thickening, digestion and dewatering characteristics of these 'fines'.
To evaluate the potential of biological phosphate removal a potential difference between the phosphate content of granules and fines is important. Research on this matter executed by TUD showed that this was not the case.

7.5 Design GSBR

7.5.1 Assumptions

Primary and post treatment

Considering the primary treatment two extreme alternatives are taken into account:

- GSBR with primary treatment including chemical dosing with extensive removal of suspended solids. Post treatment is assumed not to be required;
- GSBR with only post treatment. The post treatment consists out of removal of suspended solids from the effluent from the GSBR. For suspended solids removal the following conventional techniques are considered:
    - conventional sedimentation;
    - lamella settlers;
    - rapid sand filtration.

Configuration GSBR

From the research appeared that granules can be grown in both an airlift reactor as in a bubble column. Because of the lower building costs bubble columns are taken as a basis for the calculations.

Sludge concentration

The sludge concentration in the GSBR after effluent discharge is assumed to be 10 kg/m^3. A free space of 1 m between sludge level and discharge location is assumed.

Sludge production and characteristics

The design sludge load and sludge production in a GSBR are based on the results of the STOWA research with airlift reactors (97-25). In this research the net sludge production is zero, meaning equal sludge content in influent and effluent. A sludge production of 0.08 kgDS/kgCOD resulting from dissolved COD degradation is assumed. Total sludge production excluding chemical sludge before digestion amounts to 0.4 kgDS/kgCOD.

With regard to the sludge characteristics fines washed out from the GSBR are assumed to be removed by sedimentation. This means that biological processes in the GSBR convert colloidal particles in the influent into settlable solids. This is based on the results of the aforementioned STOWA research with airlift reactors (97-25). In this research SVI-values of fines washed out of the airlift reactors between 60 and 200 ml/g were reported. The removal and conversion of colloidal particles from the wastewater is an important point of interest in follow-up research.

Sludge treatment

The sludge treatment for the GSBR-alternatives consists of thickening, digestion and dewatering, which corresponds to the sludge treatment process of the reference alternatives.

COD- and nitrogen removal

Extensive COD- and nitrogen is assumed to be possible. The design sludge load is based on the STOWA research 97-25. High removal rates for COD and N-total were reported at sludge loads of 0.3 kgCOD/(kgDS.d) respectively 0.05 kgNkj/(kgDS.d).

The assumed removal efficiencies for COD and suspended solids are given in table 13.

Table 13 Removal efficiencies GSBR

Parameter	Suspended	Dissolved
GSBR with primary sedimentation and chemical dosing		
COD	15%	85%
Suspended solids	15%	n.a.
GSBR with post treatment		
COD	20%	85%
Suspended solids	25%	n.a.

Phosphate removal
Only biological phosphate removal is considered.

Construction height
One of the GSBR-alternatives is based on biological phosphate removal which is assumed to be possible in a full-scale GSBR applying alternating anaerobic feeding periods and aerobic reaction periods. In order to be assured of plug flow conditions while feeding from the bottom of the GSBR through the settled granules bed, the maximum hydraulic surface load during the feeding period is chosen at 7.5 m/h. This leads to a construction height of the GSBRs of 5-6 m.

Number of treatment lines
The number of parallel treatment lines is determined by the length of filling time compared to total cycle time. Assumed is that at the most one reactor can be fed with wastewater. For instance, if the filling time is a third part of the total cycle time, this would lead to three parallel treatment lines.

Effluent discharge
The effluent is assumed to be discharged by gravity from a fixed height. Furthermore a maximum flow velocity in the effluent pipe of 2.8 m/s is assumed. This is a compromise between maximum discharge flow at one hand and hydraulic losses in the effluent pipes at the other hand. A low maximum flow rate leads to large diameters, a low hydraulic decline and high discharge flows (short decantation times). Disadvantages of high discharge flows are expensive discharge constructions as well as high costs for post treatment (if required). High design discharge flow velocities result in high hydraulic losses and therefore in long decantation times.

Cycle times
For the filling time and total cycle time respectively 20 and 60 minutes are assumed. The sedimentation time is calculated based on the sedimentation velocity and the reactor height. The decantation time is calculated based on the maximum flow velocity in the effluent pipe and the reactor height.

Post treatment

Surface load conventional sedimentation	:	1	m/h
Surface load lamella settler (based on footprint)	:	8	m/h
Surface load sand filters	:	20	m/h

Alternatives
In appendix 2, figures 1 and 2, the global process flow diagrams for the GSBR alternatives are given. The alternative with post treatment is depicted with lamella settlers.

7.5.2 Buffer capacity

Raw wastewater

The total wastewater flow including rainwater is treated in the GSBR. Considering the flat RWF flow pattern (see figure 14) buffering of raw wastewater is of no use. Because buffering of wastewater is pointless at RWF-conditions, buffering raw wastewater at DWF-conditions is not considered. This means that most of the time the volume of batches to be treated will vary over the day. In principle this is not different from conventional continuous systems, especially not when the number of parallel treatment lines of the GSBR increases.

Effluent GSBR

In case of post treatment, the overflow from the GSBR is buffered. Because effluent from the GSBR is discharged by gravity, the initial flow is high. In order to reduce the dimensions of the post treatment step, the overflow of the GSBR has to be buffered.

7.5.3 Design outline

In this paragraph the design of the GSBR with post treatment (lamella settler) and biological phosphate removal is specified for a capacity of 120,000 p.e. The cycle times as well as global dimensions are given in table 14.

Table 14 Cycle times

Fase	**max. DWF**	**RWF**
Filling	20	20
Aeration	28.5	20
Sedimentation	4	10
Decantation	7.5	10
Total	60	60

Number of treatment lines	:	3	
Batch size RWF	:	1,335	m^3
Batch size max. DWF	:	528	m^3
Design capacity per GSBR	:	3,800	m^3
Volume per GSBR at RWF	:	3,435	m^3
Sludge content GSBR at RWF	:	4.6	kg/m^3
Volume GSBR at max. DWF	:	2,630	m^3
Sludge content GSBR with max. DWF	:	6.0	kg/m^3
Volume GSBR after decantation (incl. free space)	:	1,570	m^3
Surface lamella settler	:	500	m^2

From the preceding it appears that for an important part the capacity of the GSBR at RWF is determined by the size of the batches.

7.5.4 Effluent quality

In table 15 effluent qualities for both alternatives are given. The alternative with primary treatment does not meet the required effluent quality with respect to suspended solids and P-total (compare with table 9) which is caused by a too high suspended solids concentration in the pre-treated wastewater (50 mg/l). Given the starting point that suspended solids in a GSBR are not

removed, the suspended solids concentration in the effluent is too high. This also leads to increased levels of COD, N-kj and P-total.
The GSBR with post treatment meets the effluent requirements.

The effluent requirements for the alternative with pre-treatment can only be met if the suspended solids concentration in the wastewater fed to the GSBR is less than 10-30 mg/l.

Table 15 Effluent quality

Parameter	GSBR with primary sedimentation and chemical dosing	GSBR with post treatment
COD (mg/l)	80	40
Suspended solids (mg/l)	50	< 10
Nkj (mgN/l)	4.1	2.0
NO_3^- (mgN/l)	5.9	8.0
Ptotal (mg/l)	1.5	1.0

7.5.5 Costs

In the appendices 6 and 7 the dimensions and cost estimates for the two GSBR alternatives with a capacity of 120,000 p.e. are given.
Figure 17 and 18 show the investment costs and the total specific annual costs (sum of capital costs and operational costs) in relation to the capacity. For the GSBR with post treatment lamella settlers are chosen. The costs for reference alternative 1 (AS with primary sedimentation, Bio-P and sludge digestion) are given. Also the costs of the GSBR alternative with primary treatment are depicted.
As can be seen both the investment costs and the total specific annual costs of the reference alternative are the highest for the complete range. Furthermore the costs of the GSBR with post treatment are, as expected, higher than the costs of the GSBR with primary treatment.
The investments costs of the GSBRs with primary treatment respectively post treatment are on the average 29% and 12% lower than the reference alternative. The total annual costs are 16% and 6% lower.

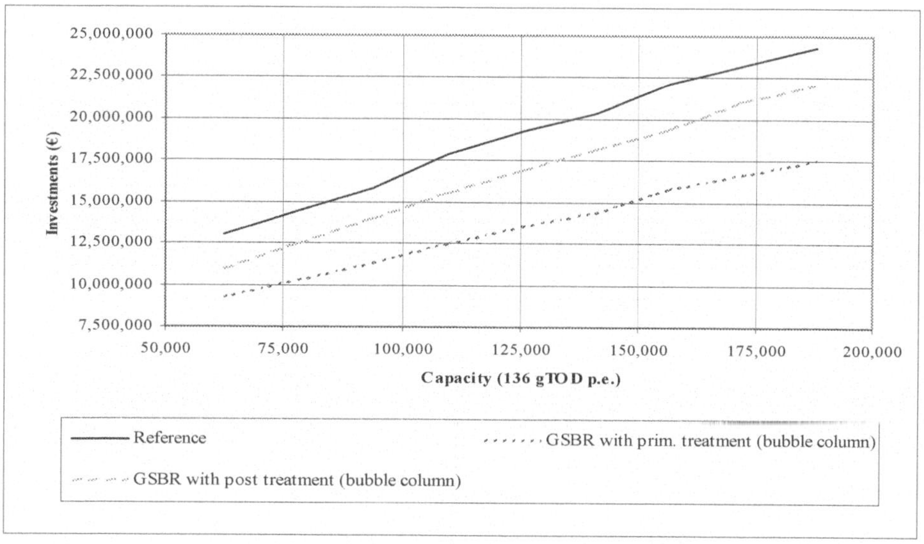

Figure 17 Investment costs

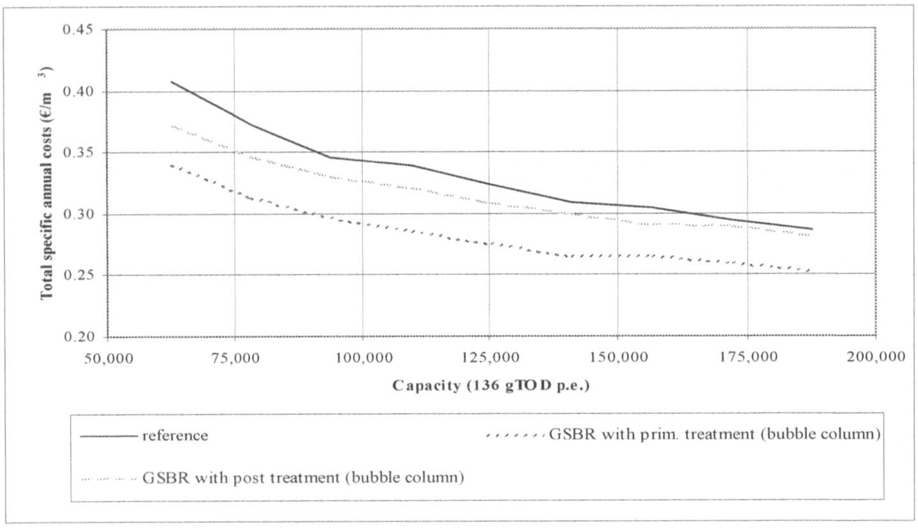

Figure 18 Total specific annual costs

7.6 Sensitivity analysis

In this paragraph a sensitivity analysis for a number of important parameters is given. For this purpose a GSBR with post treatment, biological phosphate removal and a capacity of 120,000 p.e. is used. If not mentioned otherwise, a RWF/DWF ratio of 2.5 is applied (see also paragraph 7.2.1).

7.6.1 RWF/DWF ratio

The RWF has a large impact on the design of the STP, especially for a STP based on a batch wise technology. At lower RWFs a GSBR system becomes relatively more attractive than a conventional activated sludge system. In the Netherlands a lower RWF is not imaginary given the increasing amount of improved separated sewage systems, the increasing level of disconnection of clean water and solid surface, optimisations of the wastewater chain and active control of pumping-stations (RTC). In figure 19 the influence of the RWF on the costs for a GSBR and reference alternative 1 are given. Below a RWF/DWF ratio of 3.5 a GSBR becomes more attractive.

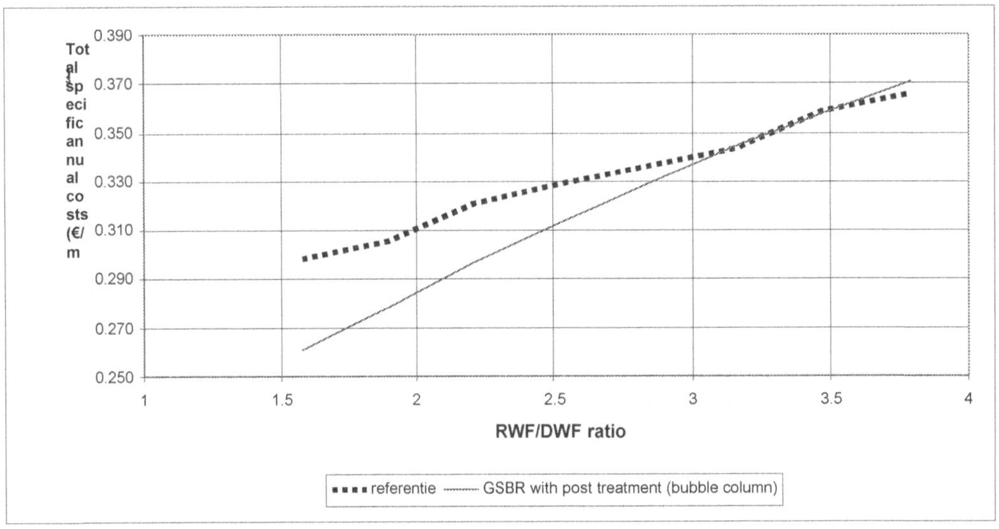

Figure 19 Influence RWF/DWF ratio

7.6.2 Land price

The sensitivity of the land price on reference alternative 1 and the GSBR alternative with post treatment is shown in figure 20. As can be seen the GSBR is more attractive on the complete range.

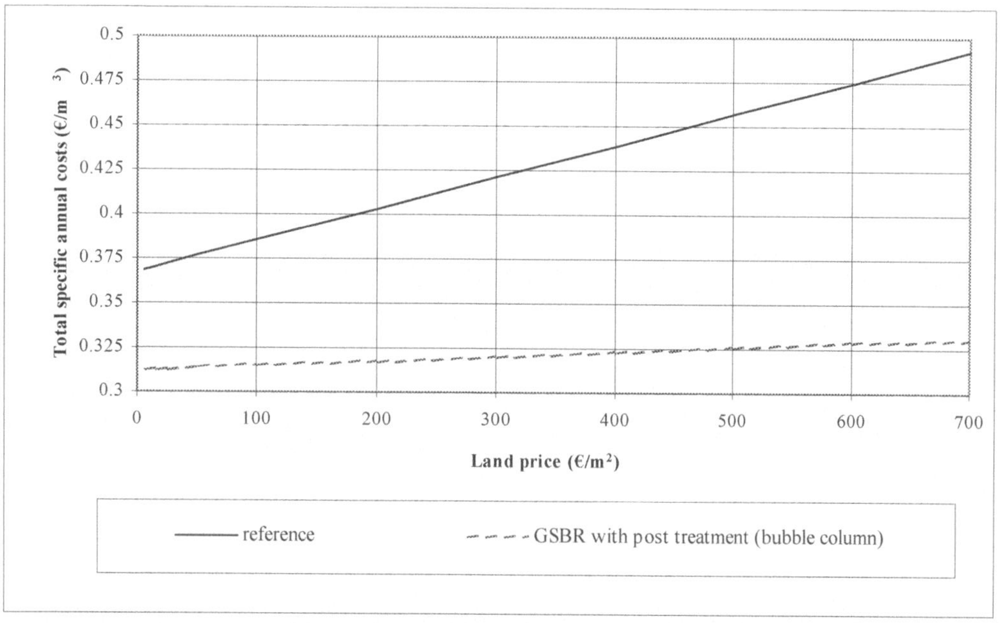

Figure 20 Influence land price

7.6.3 Post treatment alternatives

In table 16 construction costs as well as total specific annual costs of the alternatives for post treatment are given (capacity 120,000 p.e.). Based on construction costs lamella settlers are the most attractive. The total costs of a GSBR alternative with post treatment based on conventional sedimentation are the least expensive. This is because the contribution of the mechanical parts of a lamella settler to the construction costs is substantial higher compared to conventional sedimentation. Sand filtration is more expensive compared to the other two alternatives.

Table 16 Costs alternatives post treatment

Building costs	Construction costs (€)	Total costs GSBR with post treatment and Bio-P (€/m^3)
Lamella settler	1,810,000	0.313
Conventional sedimentation	1,960,000	0.310
Sand filtration	2,340,000	0.319

7.6.4 Round or rectangular?

In table 18 the basic construction costs (no constructions for the incoming wastewater and effluent disposal, no division walls, etc.) are given for round and rectangular concrete tanks with a height of 10 m. It can be concluded that round tanks are substantially cheaper in comparison with rectangular tanks.

Table 18 Basic construction costs

Capacity (m^3)	Construction costs (€/m^3)	
	Round tanks	Rectangular tanks
3.000	107	151
6.000	78	128
12.000	57	104

7.6.5 Cycle time

The influence of the cycle time is given in table 19. The cycle time appears to have a substantial influence on the batch size at RWF conditions and consequently on the volume of the GSBR. The difference in annual costs between a cycle time of 60 and 240 minutes is € 210,000,-.

Table 19 Influence cycle time

Cycle time (minutes)	Batch size RWF (m^3)	Volume GSBR (m^3)	Volume buffer tank effluent GSBR (m^3)	Total specific annual costs (€/m^3)
60	1,335	3,800	1,470	0,313
120	2,670	5,250	2,940	0,326
180	4,000	6,750	4,400	0,335
240	5,335	8,350	5,875	0,343

7.7 Conclusions and potential

7.7.1 Reference alternatives

The reference alternatives with biological phosphate removal are financially more attractive compared to the reference alternative with chemical phosphate removal. Chemical phosphate removal leads to a higher sludge production and a high chemical use. The differences between the reference alternatives 1 and 3 ("AS with primary sedimentation, Bio-P and sludge digestion" and "AS with Bio-P") are small. At lower capacities (< 150,000 p.e.) reference alternative 3 is cheaper, above 150,000 p.e. alternative 1 is more attractive.

7.7.2 GSBR

Type of primary treatment

A GSBR with only primary sedimentation and chemical dosing can not meet the required effluent standards. This is caused by an insufficient suspended solids removal efficiency in the primary treatment combined with the assumption that suspended solids are not removed in a GSBR.

A GSBR with only primary treatment can only meet the required effluent standards if the removal efficiency for suspended solids of the primary treatment is higher than 95%. This high value can be accomplished with new technologies such as membrane filtration.

During the pilot research the suspended solids removal efficiency in a GSBR will be determined. If higher suspended removal efficiencies can be established than assumed in this study, a GSBR with only primary treatment could be possible.

Under the assumptions presented in this report a GSBR with post treatment can meet the required effluent standards.

Potential
The aerobic granular sludge technology seems very promising!
Based on both the investments and the total annual costs a STP based on GSBR in the capacity range between 60,000-190,000 p.e. is economically more attractive compared to the reference alternatives based on activated sludge technology. The total annual costs for a GSBR with primary respectively post treatment are on the average 16% and 6% lower.
The footprint of a STP based on aerobic granular sludge technology amounts to only 25% compared to the references.

8 CONCLUSIONS AND RECOMMENDATIONS

The lab-scale experiments led to stable granular sludge and more insight in the important factors for simultaneous conversion of COD, P and N components from the influent. Experiments with different feeding periods, different reactor configurations and different oxygen concentrations with a synthetic influent were carried out as well as an experiment with a pre-treated sewage influent. Also the feasibility of a full-scale wastewater treatment plant with a granular sludge reactor was investigated. From this research the following conclusions can be drawn.

Long feeding times were feasible when an anaerobic feeding period is applied. A longer feeding phase is favourable because of the simplification of the scale-up and batch scheduling. Furthermore, the sequencing anaerobic feeding period and aerobic reaction period causes a domination of slow growing phosphate accumulating organisms. This last phenomenon is the main reason for stability of the granules at low dissolved oxygen concentrations. At a DO of 20% of the saturation concentration, simultaneous COD, phosphate and total nitrogen can be accomplished and effluent standards can be complied with. The use of a bubble column and thus applying lower shear was also achievable because of selection for slow growing PAOs.

Operation of a reactor with pre-treated municipal wastewater proved that granulation also occurs with complex substrate.

The feasibility study showed that the aerobic granular sludge technology seems very promising. Based on total annual costs both GSBR variants prove to be more attractive than the reference alternative (6-16%). From a sensitivity analysis it appears that the GSBR technology is less sensitive to the land price and more sensitive to a higher RWF. This means that the GSBR technology becomes more attractive at lower RWF/DWF ratios and at higher land prices. Because of the high permissible volumetric load the footprint of the GSBR variants is only 25% compared to the reference. However, the GSBR with only primary treatment can not meet the present effluent standards for municipal wastewater in the Netherlands, mainly because of a too high suspended solids concentration in the effluent.
A growing number of sewage treatment plants in the Netherlands is going to be faced with more stringent effluent standards. In general, activated sludge plants will have to be extended with a post treatment step (e.g. sand filtration) or transformed into a Membrane Bio Reactor. In this case a GSBR variant with primary treatment as well as post treatment can be an attractive alternative.

Based on both the good research results and the positive technical and financial evaluation for full scale application for sewage treatment, it is recommended to develop the aerobic sludge technology further by means of a pilot research at a sewage treatment plant.
Important aspects that need to be investigated are:

- the stability of granulation with domestic wastewater as a substrate and the influence of:
    - dynamic processes such as the wastewater flow pattern (at DWF and RWF) and process temperature;
    - the influence of suspended solids in the influent;
    - configuration of the GSBR (airlift, bubble column);
    - simultaneous fill and draw of the GSBR.
- volumetric loading rate;
- the effluent quality with respect to current and future effluent standards in the Netherlands;
- characteristics of granules and suspended solids washed out from the GSBR (settling characteristics, possibilities for digestion, influence of coagulants, granule size distribution)

Important aspects to investigate at laboratory scale are:

- granulation through slowly growing organisms others than PAOs;
- effects of COD/N-ratio;
- further research to the stability and composition of the granules;
- activity measurements, kinetics of conversion processes;
- degradation of slowly biodegradable components.

9 SYMBOLS AND ABBREVIATIONS

Bio-P	Biological phosphate removal
COD	Chemical Oxygen Demand
DO	Oxygen concentration expressed as percentage from the maximum soluble oxygen
DWF	Dry weather flow
EBPR	Enhanced Biological Phosphorus Removal
EPS	Extra cellular Polymer Structures
GAO	Glycogen accumulating organisms
GSBR	Granule Sequencing Batch Reactor – full scale design of the granule system
PAO	Phosphate accumulating organisms
PHA	Polyhydroxyalkanoaat
PHB	Poly-β-Hydroxybutyrate
PHV	Polyhydroxyvalaraat
RWF	Rain weather flow (including DWA)
SBAR	Sequenced Batch Airlift Reactor
SBBC	Sequenced Batch Bubble column
SBR	Sequenced Batch reactors
SRT	Solid Retention Time
SVI_5	Sludge Volume Index, measured after 5 minutes of settling
TOC	Total Organic Carbon
TSS	Total Suspended Solids
VSS	Volatile Suspended Solids

Chapter 7:

a	specific surface (m^2/m^3)
c^*_l	oxygen saturation concentration at half the height of the reactor and adjusted for oxygen depletion (ca. 14 mg/l)
c_l	average oxygen concentration in GSBR (2 mg/l)
k_l	mass transfer coefficient (m/h)
OC	aeration capacity (kg/h)
V	capacity GSBR (m^3)
v_{gs}	superficial gas velocity (m/s)

10 LITERATURE

Beun, J.J. (2001). PHB metabolism and N-removal in sequencing batch granular sludge reactors.Ph. D. thesis. Technical University, The Netherlands.

Beun, J.J., F. Paletta, M.C.M.V. Loosdrecht and J.J. Heijnen (2000). Stoichiometry and kinetics of poly-b-hydroxybutyrate metabolism in aerobic, slow growing, activated sludge cultures. *Biotechnology and Bioengineering* **67** (4): 379-389.

Beun, J.J., M.C.M. van Loosdrecht and J.J. Heijnen (1999). *Aerobic Granulation*. Conference on Biofilm Systems, New York.

Beun, J.J., M.C.M. Van Loosdrecht and J.J. Heijnen (2001). N-removal in a granular sludge sequencing batch airlift reactor. *Biotechnology and Bioengineering* **75** (1): 82-92.

Beun, J.J., M.C.M. van Loosdrecht and J.J. Heijnen (2002). Aerobic granulation in a sequencing batch airlift reactor. *Water Research* **36** (3): 702-712.

J.J. Beun, A. Hendriks, M.C.M. van Loosdrecht, E. Morgenroth, P.A. Wilderer and J.J. Heijnen Aerobic granulation in a sequencing batch reactor. Wat. Res. Vol. 33, No. 10, pp. 2283-2290

Beuthe, C.G. (1970). Über den Sauerstoffbedarf bei der biologischen Abwasserreinigung nach dem Belebtschlammverfahren. 111 H12: 689-693

Brdjanovic, D. (1998). Modeling Biological Phosphorus removal in Activated Sludge Systems.Proefschrift. Technische Universiteit Delft, Nederland.

Brdjanovic, D., M.C.M. van Loosdrecht , C.M. Hooijmans, G.J. Alaerts and J.J. Heijnen (1998). Minimal aerobic sludge retention time in biological phosphorus removal systems. *Biotechnology and Bioengineering* **60** (3): 326-332.

Chudoba, J. and Tucek, F. (1985). Production, degradation, and composition of activated sludge in aeration systems without primary sedimentation. *Journal WPCF* **57** (3): 210-206.

Etterer, T. and P.A. Wilderer (2001). Generation and properties of aerobic granular sludge. *Water Science & Technology* **43** (3): 19-26.

Falkentoft, C., Maria (2000). Simultaneous removal of nitrate and phosphorus in a biofilm reactor; the aspect of diffusion.Ph. D. Study. Technical University of Denmark, Denmark.

Gjaltema, A., J.L. Vinke, M.C.M. Van Loosdrecht and J.J. Heijnen (1997). Abrasion of suspended biofilm pellets in airlift reactors: Importance of shape, structure and particle concentrations. *Biotechnology and Bioengineering* **53** : 88 - 99.

Henze, M., W. Gujer, T. Mino, T. Matsuo, M.C. Wentzel, G.v.R. Marais and M.C.M. Van Loosdrecht (1999). Activated Sludge Model No.2d, ASM2D. *Water Science and Technology* **39** (1): 165 - 182.

Kayser, R. (1983). Ein Ansatz zur Bemessung einstufiger Belebungsanlagen für Nitrifikation-Denitrifikation. *GWF Wasser/Abwasser* **124** H9: 419-427

Kuba, T., G. Smolders, M.C.M. Van Loosdrecht and J.J. Heijnen (1993). Biological phosphorus removal from wastewater by anaerobic, anoxic sequencing batch reactor. *Water Science and Technology* **27** (5-6): 241-252.

Meinhold, J., H. Pedersen, E. Arnold, S. Isaacs and M. Henze (1998). Effect of continuous addition of an organic substrate to the anoxic phase on biological phosphorus removal. *Water Science and Technology* **38** (1): 97-105.

Mino, T., W.-T. Liu, F. Kurisu and T. Matsuo (1995). Modelling glycogen storage and denitrification capability of microorganisms in enhanced biological phosphate removal processes. *Water Science and Technology* **31** (2): 25-34.

NNI (1982). Bepaling van de asrest. Delft, Nederlands Normalisatie Instituut.

Okabe, S., K. Hiratia, Y. Ozawa and Y. Watanabe (1996). Spatial microbial distributions of nitrifiers and heterotrophs in mixed-population biofilms. *Biotechnology and Bioengineering* **50** (1): 24-35.

Picioreanu, C. (1999). Multidimensional modeling of biofilm structure. Proefschrift. Technische Universiteit Delft, Nederland.

Smolders, G.J.F. (1995). A metabolic model of the biological phosphorus removal: stoiciometry, kinitecs and dynamic behaviour. Proefschrift. Technische Universiteit Delft, Nederland.

Smolders, G.J.F., J. van der Meij, M.C.M. Van Loosdrecht and J.J. Heijnen (1994a). Stoiciometric model of the aerobic metabolism of the biological phosphorus removal process. *Biotechnology and Bioengineering* **44** : 837-848.

Smolders, G.J.F., J. van der Meij, M.C.M. Van Loosdrecht and J.J. Heijnen (1994b). Stoiciometry and pH influence of the anaerobic metabolism. *Biotechnology and Bioengineering* **43** : 461-470.

STOWA (1997). Zuivering van stedelijk afvalwater in een airlift-biofilmreactor - uitontwikkeling en demonstratie. Utrecht, STOWA: 96.

STOWA (2001). Handboek biologische fosfaatverwijdering. Utrecht, Stichting Toegepast Onderzoek Waterbeheer: 234.

Tijhuis, L., B. Hijman, M.C.M. Van Loosdrecht and J.J. Heijnen (1995). Influence of detachment, substrate loading and reactor scale on the formation of biofilms in airlift reactors. *Applied microbiology and biotechnology* **45** : 7-17.

Van Loosdrecht, M.C.M. (1995). Biofilm structures. *Water Science and Technology* **32** (8): 35-43.

Van Loosdrecht, M.C.M., L. Tijhuis, A.M.S. Wijdieks and J.J. Heijnen (1995). Population distribution in aerobic biofilms on small suspended particles. *Water Science and technology* **31** (1): 163 - 171.

APPENDIX 1

PFDS REFERENCE SYSTEMS

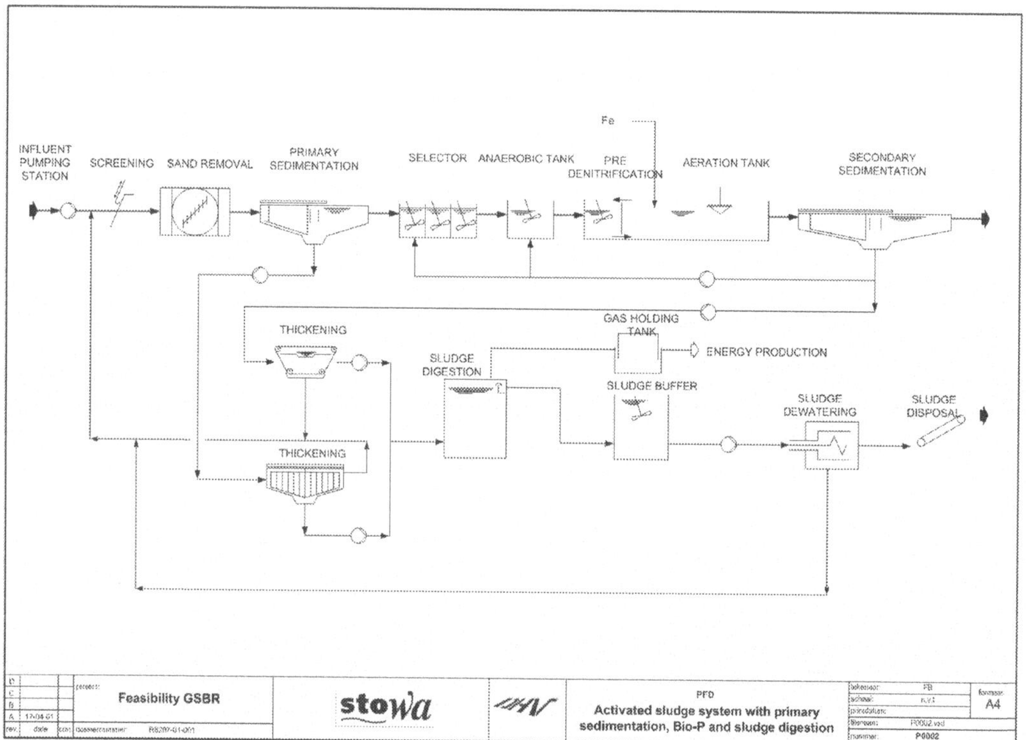

Figure 1 Process flow diagram AS with primary sedimentation, Bio-P and sludge digestion

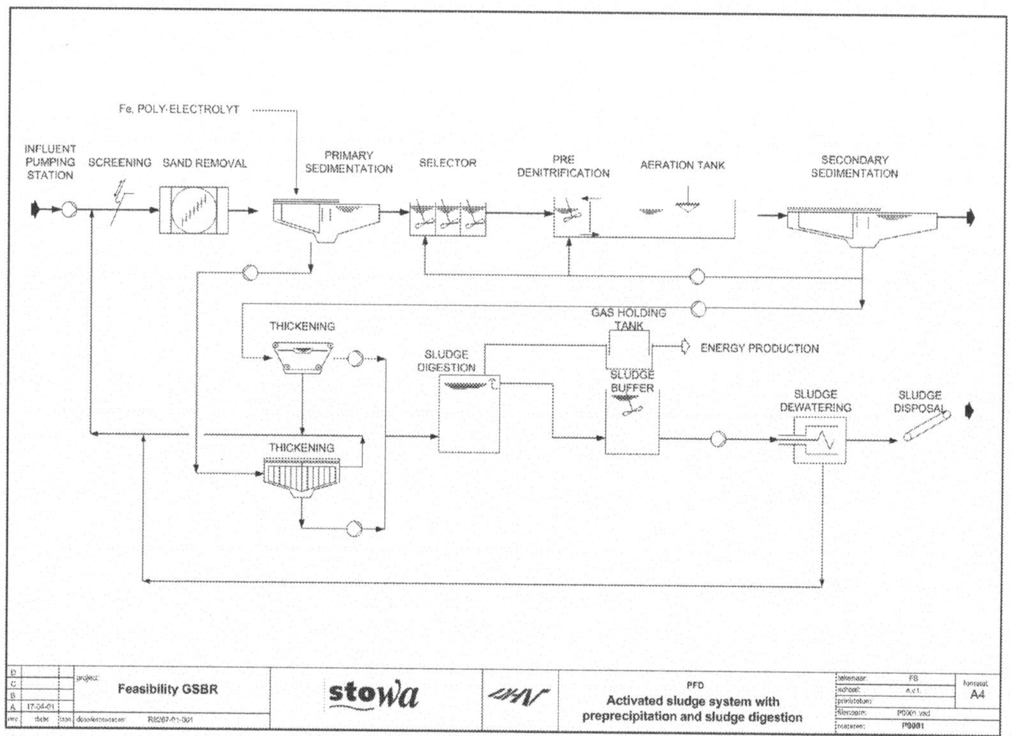

Figure 2 Process flow diagram AS with primary sedimentation, pre-precipitation and sludge digestion

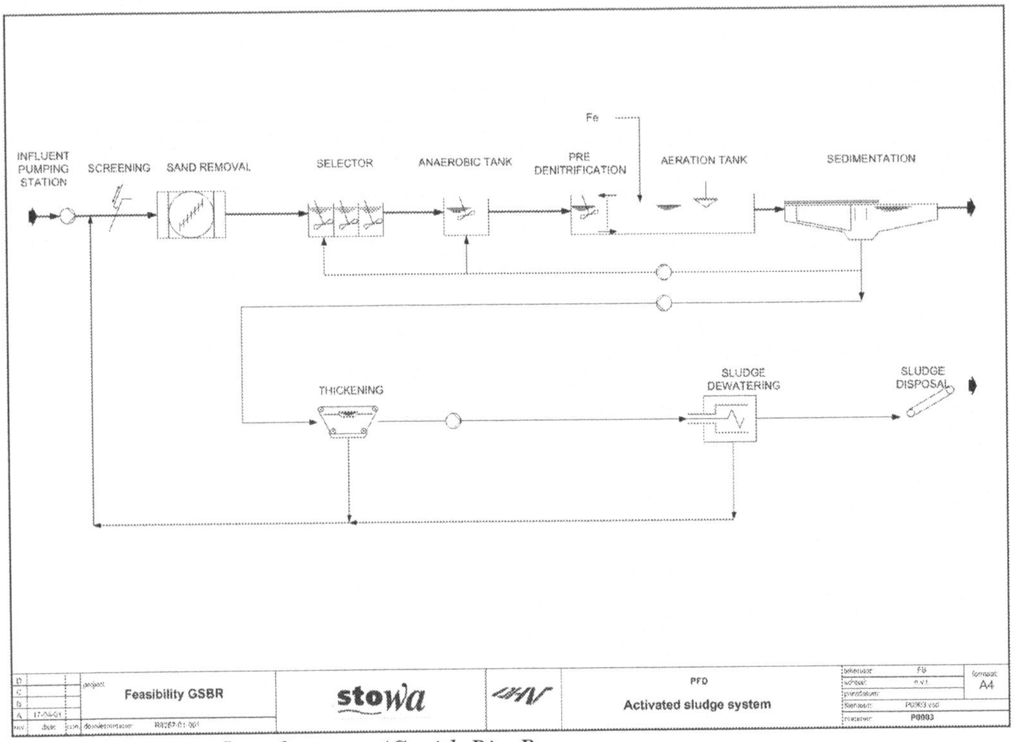

Figure 3 Process flow diagram AS with Bio-P

APPENDIX 2

PFDS GSBR ALTERNATIVES

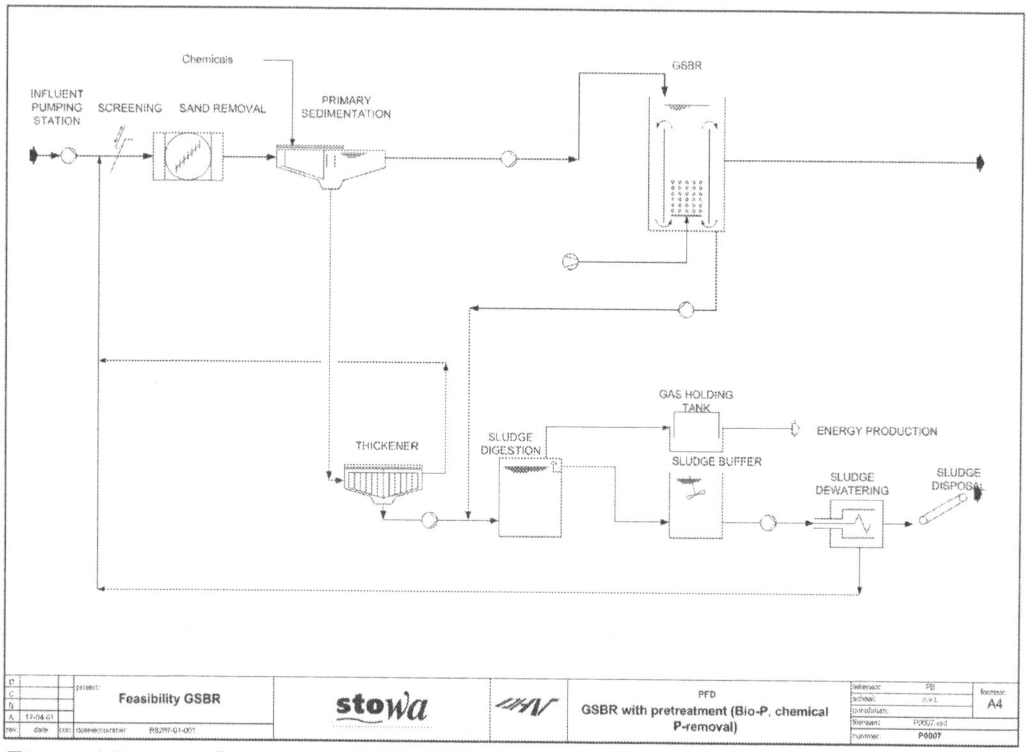

Figure 1 Process flow diagram GSBR-alternative with primary treatment

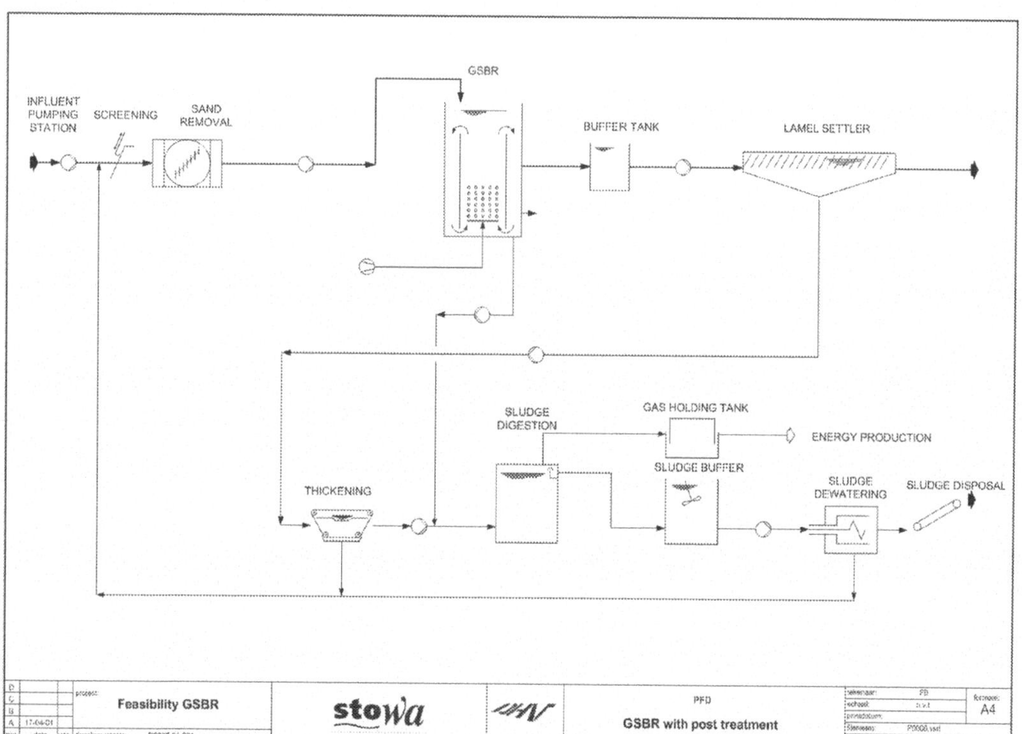

Figure 2 Process flow diagram GSBR-alternative with post treatment

APPENDIX 3

DESIGN AND COSTS OF REFERENCE 1

REFERENCE ALTERNATIVE	Activated sludge with prim. Sed., Bio-P and sludge digestion			
INFLUENT		**VALUE**	**UNIT**	
COD		600	mg/l	
BOD		220	mg/l	
Nkj		55	mg/l	
SS		250	mg/l	
Average daily flow (ADF)		**19,000**	m3/d	
Rain weather flow (RWF)		**4,000**	m3/h	
Pollution load		**118,939**	136 TOD p.e.	
EFFLUENT				
N-total		**9**	mg/l	
Operational Costs		**Consumption/Production**		**Costs (€/year)**
Sludge production		1,179	TonDS/y	377,303
Energy consumption aerators minus energy production		1,214,109	kWh/y	65,562
$FeCl_3$		327	Ton FeCl3 (40%)	37,594
Poly electrolyte		32	Ton/year	114,121
Maintenance				136,306
TOTAL				730,886
Capital costs				**Costs (€/year)**
Civil				946,861
Mechanical				416,751
Electricial				166,700
Total				1,530,312
Interest land purchase				20,992
TOTAL YEARLY COSTS				**2,282,190**
				0.33 €/m3

PROCESS UNIT	TYPE		NUMBER	CAPACITY	UNIT	CONSTRUCTION COSTS Civil (€)	Mechanical (€)
DESIGN WATER TREATMENT							
INFLUENT PUMPING STATION	PUMPS		3	1330	m3/h	204,201	183,781
SCREENINGS REMOVAL	OPENING SIZE (mm)	3	2	2000	m3/h	181,512	108,907
SAND REMOVAL (WATER LINE)	FLAT		1	100	m2	164,442	81,171
PRIMARY CLARIFIER	CIRCULAR		1	36	m	457,095	81,050
PREPRECIPITATION			NO				
	FeCl3-tank				m3		
SELECTOR	ANOXIC		1	494	m3		
MIXERS					kW		
ANAEROBIC TANK			1	3125	m3	539,545	
MIXERS	kW		4	3.9	kW		53,318
AERATION TANK	PREDENITRIFICATION		2	7092	m3	1,699,824	
AERATOR	SURFACE		4	90	kW		149,929
SIMULTANEOUS PRECIPITATION	FeCl3 -DOSING		2	25	l/h		2,503
	FeCl3-tank		1	7.0	m3		15,126
SECONDARY CLARIFIER	CIRCULAR		3	49	m	1,766,701	390,997
RETURN SLUDGE PUMP	FO		3	1011	m3/h	54,454	99,078
REJECTION WATER	RECIRCULATION			78	m3/d		
DESIGN SLUDGE TREATMENT							
PRIMARY SLUDGE PUMP			2	9	m3/h	1	7,427
PRIMARY SLUDGE THICKENER	GRAVITARY		1	6.2	m	55,778	28,468
SURPLUS SLUDGE PUMP			2	16	m3/h	2	8,814
SURPLUS SLUDGE THICKENER	MECHANICAL		2	14	m3/h	0	152,498
DIGESTION TANK			1	1808	m3	409,629	151,736
GAS STORAGE TANK			1	367	m3	0	88,982
GAS MOTOR			1	128	kW		144,805
SLUDGE BUFFER			1	271	m3	86,773	
Mixers	7	W/m3	1	1.9	kW		7,984
SLUDGE DEWATERING	CENTRIFUGE		2	10	m3/h		254,117
PE DOSING	LIQUID		2	70	l/h		119,279
DEWATERED SLUDGE PUMP			2	2	m3/h		34,363
CENTRATE PUMP			2	16	m3/h		28,044
SLUDGE STORAGE	CONTAINER		2	20	m3		9,076
PIPING	Deel civiel	8%				449,596	
OTHER	Deel civiel	6%				337,197	
BUILDINGS							
Operation building						273,000	
Sludge treatment						409,000	
Subtotal						7,088,750	2,201,452

Construction costs		**Contingency**	**Costs (€)**
Civiel		20%	7,088,750
Mechanical		20%	2,201,452
Electrical	40%	20%	880,581
Total construction costs			12,204,939
Lay-out costs		2%	244,099
Additional costs		5%	610,247
Unexpected		10%	1,220,494
Total additional costs			2,074,840
Construction + additional costs			14,279,778
Design & Engineering		10%	1,427,978
Subtotal			15,707,756
VAT		19%	2,984,474
Investments			18,700,000
Land purchase			349,865

OTHER			
Average oxgen consumption		11,137	kgO_2/d
Methane production		880	m^3/d
Footprint		15,413	m^2

APPENDIX 4

DESIGN AND COSTS OF REFERENCE 2

REFERENCE ALTERNATIVE — **Activated sludge with preprecipitation and sludge digestion**

INFLUENT	VALUE	UNIT
COD	600	mg/l
BOD	220	mg/l
Nkj	55	mg/l
SS	250	mg/l
Average daily flow (ADF)	**19,000**	m3/d
Rain weather flow (RWF)	**4,000**	m3/h
Pollution load	**118,939**	136 TOD p.e.

EFFLUENT		
N-total	**10**	mg/l

Operational Costs	Consumption/Production		Costs (€/year)	
Sludge production	1,669	TonDS/y	534,001	
Energy consumption aerators minus energy production	848,033	kWh/y	45,794	
$FeCl_3$	1,551	Ton FeCl3 (40%)	178,390	
Poly electrolyte	47	Ton/year	168,835	
Maintenance			132,313	
TOTAL			1,059,332	
Capital costs			**Costs (€/year)**	
Civil			874,786	
Mechanical			416,170	
Electricial			166,468	
Total			1,457,423	
Interest land purchase			18,399	
TOTAL YEARLY COSTS			**2,535,154**	
			0.366	€/m3

PROCESS UNIT	TYPE		NUMBER	CAPACITY	UNIT	CONSTRUCTION COSTS Civil (€)	Mechanical (€)
DESIGN WATER TREATMENT							
INFLUENT PUMPING STATION	PUMPS		3	1330	m3/h	204,201	183,781
SCREENINGS REMOVAL	OPENING SIZE (mm)	3	2	2000	m3/h	181,512	108,907
SAND REMOVAL (WATER LINE)	FLAT		1	100	m2	164,442	81,171
PRIMARY CLARIFIER	CIRCULAR		1	36	m	457,095	81,050
PREPRECIPITATION	FeCl3 -DOSING		2	140	l/h		1,815
	FeCl3-tank			25	m3		27,227
SELECTOR	ANOXIC		1	466	m3	177,465	
MIXERS			3	0.8	kW		16,594
ANAEROBIC TANK			NO				
MIXERS	kW				kW		
AERATION TANK	PREDENITRIFICATION		2	5946	m3	1,533,696	
AERATOR	SURFACE		4	75	kW		145,339
SIMULTANEOUS PRECIPITATION			NO				
	FeCl3-tank		1		m3		
SECONDARY CLARIFIER	CIRCULAR		3	47	m	1,707,332	383,059
RETURN SLUDGE PUMP	FO		3	879	m3/h	54,454	94,593
REJECTION WATER	RECIRCULATION			106	m3/d		
DESIGN SLUDGE TREATMENT							
PRIMARY SLUDGE PUMP			2	21	m3/h	2	11,436
PRIMARY SLUDGE THICKENER	GRAVITARY		1	9.2	m	83,816	32,847
SURPLUS SLUDGE PUMP			2	10	m3/h	1	7,460
SURPLUS SLUDGE THICKENER	MECHANICAL		2	8	m3/h	0	127,270
DIGESTION TANK			1	2485	m3	474,825	194,128
GAS STORAGE TANK			1	391	m3	0	90,856
GAS MOTOR			1	136	kW		154,512
SLUDGE BUFFER			1	373	m3	104,106	
Mixers	7	W/m3	1	2.6	kW		9,407
SLUDGE DEWATERING	CENTRIFUGE		2	10	m3/h		254,117
PE DOSING	LIQUID		2	70	l/h		119,279
DEWATERED SLUDGE PUMP			2	2	m3/h		34,846
CENTRATE PUMP			2	16	m3/h		28,203
SLUDGE STORAGE	CONTAINER		2	20	m3		9,076
PIPING	Deel civiel	8%				411,436	
OTHER	Deel civiel	6%				308,577	
BUILDINGS							
Operation building						273,000	
Sludge treatment						409,000	
Subtotal						6,544,959	2,196,974

Construction costs		Contingency	Costs (€)
Civiel		20%	6,544,959
Mechanical		20%	2,196,974
Electrical	40%	20%	878,790
Total construction costs			11,544,867
Lay-out costs		2%	230,897
Additional costs		5%	577,243
Unexpected		10%	1,154,487
Total additional costs			1,962,627
Construction + additional costs			13,507,495
Design & Engineering		10%	1,350,749
Subtotal			14,858,244
VAT		19%	2,823,066
Investments			17,700,000
Land purchase			306,645

OTHER		
Average oxgen consumption	9,694	kgO_2/d
Methane production	939	m^3/d
Footprint	13,509	m^2

APPENDIX 5

DESIGN AND COSTS OF REFERENCE 3

REFERENCE ALTERNATIVE	Activated sludge with Bio-P

INFLUENT	VALUE	UNIT
COD	600	mg/l
BOD	220	mg/l
Nkj	55	mg/l
SS	250	mg/l
Average daily flow (ADF)	**19,000**	m3/d
Rain weather flow (RWF)	**4,000**	m3/h
Pollution load	**118,939**	136 TOD p.e.

EFFLUENT	VALUE	UNIT
N-total	**10**	mg/l

Operational Costs	Consumption/Production		Costs (€/year)
Sludge production	1,658	TonDS/y	530,492
Energy consumption aerators minus energy production	2,595,418	kWh/y	140,153
FeCl3	411	Ton FeCl3 (40%)	47,310
Poly electrolyte	46	Ton/year	164,547
Maintenance			107,249
TOTAL			989,751
Capital costs			**Costs (€/year)**
Civil			790,754
Mechanical			317,376
Electricial			126,950
Total			1,235,080
Interest land purchase			19,781
TOTAL YEARLY COSTS			**2,244,612**
			0.32 €/m3

PROCESS UNIT	TYPE		NUMBER	CAPACITY	UNIT	CONSTRUCTION COSTS Civil (€)	Mechanical (€)
DESIGN WATER TREATMENT							
INFLUENT PUMPING STATION	PUMPS		3	1330	m3/h	204,201	183,781
SCREENINGS REMOVAL	OPENING SIZE (mm)	3	2	2000	m3/h	181,512	108,907
SAND REMOVAL (WATER LINE)	FLAT		1	100	m2	164,442	81,171
PRIMARY CLARIFIER			NO				
PREPRECIPITATION			NO				
	FeCl3-tank				m3		
SELECTOR	ANOXIC		1	534	m3		
MIXERS					kW		
ANAEROBIC TANK			1	2632	m3	476,385	
MIXERS	kW		3	4.4	kW		44,353
AERATION TANK	PREDENITRIFICATION		2	8834	m3	1,871,714	
AERATOR	SURFACE		4	90	kW		149,929
SIMULTANEOUS PRECIPITATION	FeCl3 -DOSING		2	30	l/h		2,666
	FeCl3-tank		1	9.0	m3		16,639
SECONDARY CLARIFIER	CIRCULAR		3	48	m	1,714,106	383,965
RETURN SLUDGE PUMP	FO		3	893	m3/h	54,454	95,059
REJECTION WATER	RECIRCULATION			58	m3/d		
DESIGN SLUDGE TREATMENT							
PRIMARY SLUDGE PUMP			NO				
PRIMARY SLUDGE THICKENER			NO				
SURPLUS SLUDGE PUMP			2	27	m3/h	2	14,280
SURPLUS SLUDGE THICKENER	MECHANICAL		2	22	m3/h	0	180,521
DIGESTION TANK			NO				
GAS STORAGE TANK			NO				
GAS MOTOR			NO				
SLUDGE BUFFER			1	227	m3	77,119	
Mixers		7 W/m3	1	1.6	kW		7,423
SLUDGE DEWATERING	CENTRIFUGE		2	5	m3/h		226,890
PE DOSING	LIQUID		2	60	l/h		114,093
DEWATERED SLUDGE PUMP			2	2	m3/h		31,765
CENTRATE PUMP			2	10	m3/h		22,607
SLUDGE STORAGE	CONTAINER		2	20	m3		9,076
PIPING	Deel civiel	8%				379,515	
OTHER	Deel civiel	6%				284,636	
BUILDINGS							
Operation building						227,000	
Sludge treatment						273,000	
Subtotal						5,908,085	1,673,125

Construction costs		Contingency	Costs (€)
Civiel		20%	5,908,085
Mechanical		20%	1,673,125
Electrical	40%	20%	669,250
Total construction costs			9,900,551
Lay-out costs		2%	198,011
Additional costs		5%	495,028
Unexpected		10%	990,055
Total additional costs			1,683,094
Construction + additional costs			11,583,645
Design & Engineering		10%	1,158,364
Subtotal			12,742,009
VAT		19%	2,420,982
Investments			15,200,000

Land purchase	329,681

OTHER		
Average oxgen consumption	12,698	kgO_2/d
Methane production	0	m^3/d
Footprint	14,523	m^2

APPENDIX 6

DESIGN AND COSTS GSBR WITH POST TREATMENT AND BIO-P

CONFIGURATION OVERVIEW	
P removal	Biological
Pretreatment	No pretreatment
Cycle time	60 min
Reactor type	Bubble column
Buffer effluent GSB	yes
Post treatment	Lamella settlers
Discharge method	Separate fill and draw GSBR

INFLUENT	Value	Unit
COD	600	mg/l
BOD	220	mg/l
Nkj	55	mg/l
SS	250	mg/l
Ptotal	9.0	mg/l
Average daily flow (ADF)	19,000	m3/d
Rain weather flow (RWF)	4,006	m3/h
Dry weather flow (DWF)	1,583	m3/h
Pollution load	119,000	136 gTOD p.e.

EFFLUENT	Concentration	
COD	43	mg/l
SS	10	mg/l
N-sludge	0.5	mg/l
NH4	1.5	mg/l
NO3	8.0	mg/l
Ntotal	10.0	mg/l
Ptotal	0.9	mg/l

OPERATIONAL COSTS	Consumption/Production		Costs	
Sludge production	1,098	ton DS/j	351,383	€/year
Energy use airlift and other main users minus	1,984,722	kWh/j	107,175	€/year
$FeCl_3$	0	ton/j	0	€/year
Poly electrolyte	22	ton/j	79,061	€/year
Maintenance			166,391	€/year
TOTAL			704,010	€/year
Capital costs			**Costs**	
Civil			536,366	€/year
Mechanical			663,155	€/year
Electrical			265,262	€/year
Total			1,464,783	€/year
Interest land purchase			4,133	€/year
TOTAL YEARLY COSTS			**2,170,000**	€/year
			0.313	€/m3

PROCESS UNIT		NUMBER	CAPACITY	UNIT	CONSTRUCTION COSTS Civil	CONSTRUCTION COSTS Mechanical
Influent pumping station		3	1,335	m3/h	204,600	184,140
Screens		2	2,003	m3/h	181,586	108,981
Sand removal		1	100	m2	164,473	81,224
Primary clarifier	no	FALSE	FALSE	m	FALSE	FALSE
FeCl3-dosing		FALSE	FALSE	l/h		FALSE
FeCl3-tank		FALSE	FALSE	m3		FALSE
Pumping station GSBRs		3	1,335	m3/h	204,600	184,140
GSBRs	Bubble column					
	round	3	3,779	m3	1,128,101	
	FALSE		FALSE	m2	FALSE	
Compressors		3	6,121	m3/h		242,816
Diffusors		3	6,121	m3/h		91,663
Discharge construction		3	5,847	m3/h	0	446,209
Effluent buffer tank	yes	1	1,469	m3	162,930	
Effluent pumping station		3	1,335	m3/h	204,600	184,140
Lamella settlers	yes	1	4,006	m3/h	274,476	876,265
Conventional sedimentatio	no	FALSE	FALSE	m	FALSE	FALSE
Hydrocyclones	no	FALSE	FALSE	m3/h		FALSE
Sand filtration	no	FALSE	FALSE	m	FALSE	FALSE
Flush pump + blower		FALSE	FALSE	m3/h		FALSE
Clean water tank		FALSE	136	m3		FALSE
Primary sludge thickener	no	FALSE	FALSE	m	FALSE	FALSE
Primary sludge pump		FALSE	FALSE	m3/h		FALSE
Surplus sludge pump	yes	2	35	m3/h		22,743
Mechanical thickening	yes	2	35	m3/h		197,149
Digestion tank		1	1,816	m3	410,364	152,124
Gas holding tank		1	326	m3		85,850
Gas engine		1	113	kW		128,593
Sludge holding tank		1	272	m3		87,009
Mixers sludge holding tank		1	2			8,000
Sludge dewatering		2	10	m3/h		254,117
PE dosing sludge dewatering		1				113,445
Centrate pump		2	11	m3/h		23,916
Dewatered sludge pump		2	2	m3/h		31,803
Sludge containers		2				9,076
Piping		8%			234,858	
Other		6%			176,144	
BUILDINGS						
Operation building					272,268	
Sludge treatment					408,402	
Subtotal					4,027,404	3,513,405

Construction costs		Contingency	Costs (€)
Civil		20%	4,027,404
Mechanical		20%	3,513,405
Electrical	40%	20%	1,405,362
Total construction costs			10,735,405
Lay-out		2%	214,708
Additional costs		5%	536,770
Unexpected		10%	1,073,541
Subtotal			1,825,019
Construction + additional costs			12,560,424
Design & Engineering		10%	1,256,042
Subtotal			13,816,466
VAT		19%	2,625,129
Investments			16,400,000
Land purchase			68,889

OTHER		
Average oxygen consumption		kgO_2/d
Methane production	782	m^3/d
Footprint	3,035	m^2

APPENDIX 7

DESIGN AND COSTS GSBR WITH PRIMARY SEDIMENTATION AND BIO-P

CONFIGURATION OVERVIEW	
P removal	Biological
Pretreatment	Primary sedimentation + polymer dosing
Cycle time	60 min
Reactor type	Bubble column
Buffer effluent GSB	yes
Post treatment	No post treatment
Discharge method	Separate fill and draw GSBR

INFLUENT	Value	Unit
COD	600	mg/l
BOD	220	mg/l
Nkj	55	mg/l
SS	250	mg/l
Ptotal	9.0	mg/l
Average daily flow (ADF)	19,000	m3/d
Rain weather flow (RWF)	4,006	m3/h
Dry weather flow (DWF)	1,583	m3/h
Pollution load	119,000	136 gTOD p.e.

EFFLUENT	Concentration	
COD	85	mg/l
SS	57	mg/l
N-sludge	4.0	mg/l
NH4	1.5	mg/l
NO3	4.5	mg/l
Ntotal	10.0	mg/l
Ptotal	1.6	mg/l

OPERATIONAL COSTS	Consumption/Production		Costs	
Sludge production	1,182	ton DS/j	378,311	€/year
Energy use airlift and other main users minus	1,757,986	kWh/j	94,931	€/year
$FeCl_3$	0	ton/j	0	€/year
Poly electrolyte	51	ton/j	184,984	€/year
Maintenance			122,331	€/year
TOTAL			780,558	€/year
Capital costs			**Costs**	
Civil			492,796	€/year
Mechanical			464,564	€/year
Electrical			185,826	€/year
Total			1,143,186	€/year
Interest land purchase			3,648	€/year
TOTAL YEARLY COSTS			**1,930,000**	€/year
			0.278	€/m3

PROCESS UNIT		NUMBER	CAPACITY	UNIT	CONSTRUCTION COSTS Civil	CONSTRUCTION COSTS Mechanical
Influent pumping station		3	1,335	m3/h	204,600	184,140
Screens		2	2,003	m3/h	181,586	108,981
Sand removal		1	100	m2	164,473	81,224
Primary clarifier	yes	1	36	m	464,002	81,390
FeCl3-dosing		FALSE	FALSE	l/h		FALSE
FeCl3-tank		FALSE	FALSE	m3		FALSE
Pumping station GSBRs		3	1,335	m3/h	204,600	184,140
GSBRs	Bubble column					
	round	3	2,866	m3	924,459	
	FALSE		FALSE	m2	FALSE	
Compressors		3	7,030	m3/h		276,259
Diffusors		3	7,030	m3/h		105,274
Discharge construction		3	5,847	m3/h	0	446,209
Effluent buffer tank	no	FALSE	FALSE	m3	FALSE	
Effluent pumping station		FALSE	FALSE	m3/h	FALSE	FALSE
Lamella settlers	no	FALSE	FALSE	m3/h	FALSE	FALSE
Conventional sedimentati	no	FALSE	FALSE	m	FALSE	FALSE
Hydrocyclones	no	FALSE	FALSE	m3/h		FALSE
Sand filtration	no	FALSE	FALSE	m	FALSE	FALSE
Flush pump + blower		FALSE	FALSE	m3/h		FALSE
Clean water tank		FALSE	136	m3		FALSE
Primary sludge thickener	yes	1	8	m	72,696	31,110
Primary sludge pump		2	24	m3/h		20,555
Surplus sludge pump	no	FALSE	FALSE	m3/h		FALSE
Mechanical thickening	no	FALSE	FALSE	m3/h		FALSE
Digestion tank		1	1,956	m3	421,718	158,923
Gas holding tank		1	351	m3		87,754
Gas engine		1	122	kW		138,448
Sludge holding tank		1	65	m3		23,478
Mixers sludge holding tank		1	0			4,915
Sludge dewatering		2	5	m3/h		226,890
PE dosing sludge dewatering		2				226,890
Centrate pump		2	12	m3/h		24,706
Dewatered sludge pump		2	2	m3/h		32,849
Sludge containers		2				9,076
Piping		8%			211,051	
Other		6%			158,288	
BUILDINGS						
Operation building					272,268	
Sludge treatment					408,402	
Subtotal					3,688,144	2,453,212

Construction costs		Contingency	Costs (€)
Civil		20%	3,688,144
Mechanical		20%	2,453,212
Electrical	40%	20%	981,285
Total construction costs			8,547,169
Lay-out		2%	170,943
Additional costs		5%	427,358
Unexpected		10%	854,717
Subtotal			1,453,019
Construction + additional costs			10,000,188
Design & Engineering		10%	1,000,019
Subtotal			11,000,207
VAT		19%	2,090,039
Investments			13,100,000
Land purchase			60,806

OTHER		
Average oxygen consumption		kgO_2/d
Methane production	842	m^3/d
Footprint	2,679	m^2

IWA Publishing's authorised EU representative for General Product Safety Regulations is Diane D'Arras, 15 rue Duret, 75116 Paris, France, e-mail: safety@iwap.co.uk.

Printed and bound by CPI Group (UK) Ltd, Croydon, CR0 4YY
12/05/2026
02108868-0020